T0321040

Sustainable Construction Materials

Recycling of Spent Garnet

Sustainable Construction Materials

Recycling of Spent Garnet

Dr. Habeeb Lateef Muttashar
University Technology Malaysia
University of Imam Jafar Sadiq, Maysan Branch

CRC Press is an imprint of the
Taylor & Francis Group, an **informa** business

CRC Press
Taylor & Francis Group
6000 Broken Sound Parkway NW, Suite 300
Boca Raton, FL 33487-2742

Printed on acid-free paper

International Standard Book Number-13: 978-0-367-00227-5 (Hardback)

Library of Congress Cataloging-in-Publication Data

Names: Mubasher Al Zubaidi, Habeeb Lateef, author.
Title: Sustainable construction materials : recycled spent garnet / Habeeb
Lateef Muttashar Alzuabidi.
Description: Boca Raton : Taylor & Francis, a CRC title, part of the Taylor &
Francis imprint, a member of the Taylor & Francis Group, the academic
division of T&F Informa, plc, 2019. | Includes bibliographical references.
Identifiers: LCCN 2018035641 | ISBN 9780367002275 (hardback : acid-free paper)
Subjects: LCSH: Garnet. | Building materials.
Classification: LCC TN997.G3 M83 2019 | DDC 624.1/898--dc23
LC record available at https://lccn.loc.gov/2018035641

Visit the Taylor & Francis Web site at
http://www.taylorandfrancis.com

and the CRC Press Web site at
http://www.crcpress.com

Printed and bound in Great Britain by
TJ International Ltd, Padstow, Cornwall

I dedicated this work to:

My mother, Hajja Qismah Mezher, for her sacrifice

My father, Alhaji Lateef Muttashar Alzuabidi,
for his support and encouragement.

Contents

List of Figures

List of Tables

List of Tables

Acknowledgments

In the name of God, the Beneficent, the Merciful who has created mankind with knowledge, wisdom, and power. I would like to express my thanks to Almighty God on the successful achievement of this book.

Preface

Certainly the use of spent garnet in self-compacting geopolymer concrete as an alternative to river sand is beneficial in terms of avoiding environmental pollution and the over-exploitation of natural resources. Most of the recycling problems that limit waste disposal can be overcome by properly using spent garnet in making self-compacting geopolymer concrete. The use of spent garnet directly contributes to sustainable development and is a cost-effective way of manufacturing self-compacting geopolymer concrete and preserving natural sand from further degradation. Currently, enormous amounts of spent garnet are regularly disposed of and used for landfill, which has high transportation costs and is labor intensive. This not only pollutes the environment but precludes monetary gain. The present work will solve these existing problems by demonstrating how to systematically incorporate spent garnet in place of sand to prepare new compositions of sustainable self-compacting geopolymer concrete.

About the Author

Dr. Habeeb Lateef Muttashar is a researcher and consultant in the area of civil engineering at the University of Technology, Malaysia. Dr. Habeeb has conducted studies of geopolymer concrete, self-compacting concrete, recycling waste materials, green technology, and high-volume biomass mortar. He has an MS and a PhD in civil engineering from the University of Technology, Malaysia. For the past 6 years he has been working in the area of structures and materials. His 10-year career includes numerous citations for excellence in leadership relating to various government, industry, labor, and academia partnerships in civil engineering.

1 Performance of Self-Compacting Geopolymer Concrete Using Spent Garnet as Sand Replacement

INTRODUCTION

Rapid industrial growth has witnessed the ever-increasing utilization of sand from rivers for various construction purposes, which has led to the over-exploitation of riverbeds and disturbed the ecosystem. Numerous problems have emerged, including increases in riverbed depth, the lowering of water tables, increases in salinity, and the destruction of river embankments (Gourley, 2003). Recently, intensive research has proven that obtaining modified concrete by incorporating waste materials can lead to sustainable product development. Such concrete structures not only allow for greener and more environmentally sound construction but also protect against the excessive consumption of natural non-renewable fine aggregates (Temuujin, 2010).

Thus, the proper use of fine aggregates as alternative materials in concrete has become an absolute necessity for the replacement of river sand. In this regard, the utilization of spent garnets has emerged as a promising alternative in its own right. *Garnet* is a generic word that refers to an assemblage of multifaceted minerals of silicate compounds containing calcium (Ca), magnesium (Mg), ferrous iron (Fe) or manganese (Mn), aluminum (Al), chromium (Cr), ferric iron (Fe), or even titanium (Ti), each having analogous crystal lattice structures and varied chemical formulas (Castel, 2010). Interestingly, the angular fractures and hardness properties of garnets together with their ability to be recycled make them advantageous for numerous abrasive applications. The common chemical composition of garnet is $A_3B_2(SiO_4)_3$, wherein the element A may be Ca, Mg, ferrous iron or Mn, Al, Cr, ferric iron, or Ti (Rodina, 2013). Garnets have major industrial uses such as water jet cutting, abrasive blasting, water filtration, and others (Lindtner, 2014).

A comprehensive assessment of a major shipyard industry in the southern province of Malaysia revealed that the country imported approximately 2000 MT of garnets in 2013 alone, and a large quantity was dumped as waste. Generally, abrasive blasting is used to prepare the surfaces for coating and painting (Roskill Information Services, 2000). This technique is used for the construction of vessels, ship maintenance, and repair activities. Thus, the blasting process creates large quantities of exhausted garnet waste mixed with surface elements such as paint chips and oil. Such waste causes many environmental and health hazards such as water contamination when it enters the waterways during floods or through runoffs. Therefore, spent garnets pose a threat to the ecological balance and biodiversity. Garnets can be reused about three to five times while keeping their overall properties intact. Moreover, these recycled garnets degrade at a level beyond which they are non-reusable for abrasive blast purposes. Afterward, these inoperative garnets are removed from the shipyard and designated as *spent garnets* (Garnett, 2013). Recently, it has been recognized that the utilization of these spent garnets as a replacement for fine aggregates in self-compacting geopolymer concrete (SCGPC) may provide greener alternative construction materials to ordinary Portland cement (OPC)-based concrete.

Universally, OPC's excellent mechanical properties and moderately cheap and easy accessibility make it the most commonly used binder for the production of construction materials. Thus, OPC-based concrete is preferred for diverse purposes (Davidovits, 1991). Nonetheless, OPC manufacturing leads to the depletion of natural habitats, the manufacturing of fossil fuels, and substantially higher CO_2 emissions, which our planet can no longer afford. To overcome those threats, many dedicated efforts have been made to search for efficient alternative substances such as alkali-activated materials interpreted as geopolymers (GPs). The cost production of GP concrete is 1.7% higher than OPC for the same grade (Tahri et al., 2017).

These alternative substances are proven to be advantageous for sustainable development when industrial by-products are partially applied as precursor matter as a substitute for the main raw mineral binder, including OPC. Moreover, the final product exhibits improved characteristics over OPC-based concrete, depending on the implemented raw minerals and alkali activations. Factors such as the low heat of hydration, the rapid development of early strength, the formation of a stronger aggregate-to-matrix interface, poorer thermal conduction (TC), and elevated resistance to acid and fire (Provis, 2010) also considerably influence the overall properties of the ultimate products. Generally, alkali-activated materials are classified into two categories: (1) a high calcium system with granulated blast furnace slag as the usual precursor, where gel of calcium alumina silicate hydrate is the major product of reaction, (2) a low calcium product having class-F fly ash (FA) and metakaolin as the constituent raw materials, where gel of sodium alumina silicate hydrate in the form of a three-dimensional network is produced as the main product of reaction.

Categorically, the ability of self-compacting concretes (SCCs) to flow under their own weight without requiring any exterior compaction vibration has modernized the placement of concretes. A group of researchers from Japan first introduced the concept of SCC in the late 1980s (Domone, 2006). It was established that greatly workable concrete such as SCCs display a flow under their own weight via constrained segments in the absence of any segregation or bleeding. Such concretes must possess a comparatively small yield to guarantee enhanced flow capacity and reasonable viscosity to oppose separation and bleeding. Furthermore, they must retain homogeneity during transport, placement, and curing to guarantee sufficient structure performance and long-standing endurance.

Despite the many studies toward sand replacements for concrete infrastructures, the exploitation of spent garnet waste as construction material is seldom addressed.

Considering the notable engineering properties of spent garnet waste, this book explores the feasibility of incorporating different levels of spent garnet as a replacement for river sand to achieve an enhanced SCGPC. SCGPC specimens were thoroughly characterized to determine their compressive, flexural, workable durability and microstructures as a function of varying percentages of spent garnet inclusion.

There are three main research questions in this book:

1. What are the effects of spent garnet on the fresh and hardened characteristics of the SCGPC concrete in terms of workability and mechanical strength?
2. What are the effects of spent garnet SCGPC durability such us carbonation, sulfate attack, acid attack?
3. What are the effects of spent garnet on the morphology of SCGPC concrete such us bonding and thermal analysis?

Several experiments (for synthesis, characterization, and performance evaluation) were conducted for this work, the main focus of which was to develop a sustainable spent garnet-based SCGPC with varying levels (25%, 50%, 75%, and 100%) of replacement of river sand. The properties of the constituent concrete materials, including leaching behavior, carbonation, and thermal and mechanical characteristics, and the microstructures of the garnet were examined. The workability, mechanical strengths, deformation (modulus of elasticity), and durability characteristics of the developed SCGPC were evaluated for comparison with that of traditional concretes. Tests such as slump flow, L-box, V-funnel, T50, compressive strength, flexural strength, indirect tensile strength, drying shrinkage, modulus of elasticity, carbonation, and acid and sulfate resistance were carried out to determine the performance of formulated SCGPC. Hardened SCGPC of optimum composition was selected to examine its crystallinity, microstructure, and bonding and thermal properties using X-ray diffraction (XRD), field emission scanning electron microscopy (FESEM), Fourier-transform infrared spectroscopy (FTIR), and thermogravimetric analysis/differential thermal analysis (TGA/DTA).

SIGNIFICANCE OF THE BOOK

Certainly, the use of spent garnet in SCGPC as an alternative to river sand is beneficial in terms of avoiding environmental pollution and the over-exploitation of natural resources. Most of the recycling problems that limit waste disposal can be overcome by properly using spent garnet in making SCGPC (Lottermoser, 2011). The use of spent garnet directly contributes to sustainable development and is a cost-effective way of manufacturing SCGPC and preserving natural sand from further degradation. Currently, enormous amounts of spent garnet are regularly disposed of and used for landfill, which has high transportation costs and is labor intensive. This not only pollutes the environment but precludes monetary gain. The present work will solve these existing problems by systematically incorporating spent garnet in place of sand to prepare new compositions of sustainable SCGPC. This kind of SCGPC will be economically viable because of its high abundance, its non-toxic nature, and the cost-effectiveness of using spent garnet as the main constituent. It will be demonstrated that spent garnet is a potential substitute material for river sand in building and structural engineering. Thus, the use of spent garnet in place of fine aggregates to make concrete will avoid the over-use of natural sand. This research is expected to modernize the Malaysian construction industries and encourage builders and engineers to use eco-friendly spent garnet-based SCGPC rather than that based on conventional natural river sand.

BOOK ORGANIZATION

The present book is composed of seven chapters, as follows:

Chapter 1 provides a brief background and overview of the research to identify the research gap and clarifies the problem statement and the rationale of the research. Based on the problem to be solved, it sets a goal and relevant objectives. Furthermore, it discusses the research scope and significance.

Chapter 2 presents a comprehensive literature review to justify the problem statement. It emphasizes past developments in SCGPC, ongoing activities in the field of spent garnet-based concrete production, and future trends in SCGPC based on spent garnets as a replacement for river sand.

Chapter 3 presents the experimental results in terms of analyses, discussions, evaluations, and comparisons with other works on similar SCGPCs. The physicochemical properties of spent garnets and their effects on the properties of fresh as well as hardened SCGPC are highlighted. Results are obtained on workability using tests such as slump flow, L-box, V-funnel, and T50, and hardened properties are discussed in terms of compressive, flexural, and tensile strengths, drying shrinkage, and modulus of elasticity.

Chapter 4 explains and discusses the results obtained from different tests on durability performed on control specimens as well as spent garnet-based GP concrete. Results from durability tests on SCGPC, such as drying shrinkage, water absorption, accelerated carbonation, and resistance to acid and sulfate attacks, are investigated.

Chapter 5 presents the thermal properties, bonding vibrations, crystalline structures, surface morphology, and microstructures of spent garnets obtained via XRD, FESEM, FTIR, and TGA/DTA analysis. Furthermore, results from microstructural studies of SCGPC are presented at 6 months of strength development and above.

Chapter 6 concludes on the overall performance of spent garnet as a replacement for sand in SCGPC, its major contributions, and the novelties of the present subject.

2 Literature Review

INTRODUCTION

In recent times, the steady escalation in the amount of waste generated from mining and diversified industries has created environmental concern related to the depletion of natural resources and landfill-mediated pollution (Siddique, 2011). For the remediation of such hazards, the effective management of solid waste has become inevitable worldwide. Due to insufficient space for landfill and the increasing costs of land disposal, it has become imperative to recycle and reutilize industrial waste materials. Mine waste materials and industrial by-products are categorized into various groups. The use of these waste materials in concrete not only reduces disposal concerns and eases environmental concerns but also makes these concretes economical as well as sustainable (Singh, 2013). Natural sand is declining due to its enormous consumption in the construction industries. Therefore, it is vital to replace natural river sand with other fine aggregates that are suitable for making sustainable self-compacting geopolymer concrete (SCGPC), such as spent garnet. This is the recurring theme of this book.

In recent times, the use of spent garnet as an alternative fine aggregate to natural sand in SCGPC has become a new trend for large-scale recyclable waste. Limited literature exists on waste garnet as appropriate construction material. Diverse by-products and wastes from different industries have previously been used in concrete as fine aggregate replacements (Aggarwal, 2014). These include waste foundry sand, coal bottom ash, stone dust, recycled fine aggregate, glass cullet, and copper tailings. Many studies have been conducted to achieve optimum mix compositions with good mechanical properties. Despite many research efforts, an optimum composition of sustainable SCGPC with the desired properties based on waste materials as a suitable replacement for river sand is far from being achieved (Chan, 2013).

This chapter describes the detailed literature on waste garnet-based SCGPC production, the basic need for the development of SCGPC using spent garnet as an alternative fine aggregate to sand, and the past research activities related to waste garnets. Through a comprehensive literature survey, it is demonstrated that there are several open avenues and unsolved problems in spent garnet-based SCGPC production and optimization, the thorough characterization of samples, the control of various properties of such concretes for enhanced applications, and the remediation of environmental issues.

CONCRETES AND ENVIRONMENT

To meet the escalating demand for construction material, the concrete industries worldwide began to exploit natural resources. This over-exploitation of resources and the large-scale production of concrete constituents led to several environmental issues, including effluence, the dumping of toxic wastes, greenhouse gas emissions, climate change, global warming, and the depletion of natural resources. Lately, it has been realized that it is necessary to wean the industry off natural river sand in the production of concretes. Industrial nations all over the world are now focusing more on sustainable development by reducing greenhouse gas emissions related to global warming. Attention is being paid to producing cements with low carbon dioxide (CO_2) emissions in the replacement of ordinary Portland cement (OPC). Careful estimates show that for each ton of cement production via the calcination process of limestone and alumino-silicates, about 0.9 tons of CO_2 is discharged into the atmosphere (Davidovits, 2005). In developed countries (particularly China, India, South Korea, Taiwan, Indonesia, etc.), with the gamut of industrial expansion, the manufacturing of cement is exponentially increasing (Davidovits, 2002). According to the proposed Kyoto Protocol (1997), CO_2 emissions and PVC manufacturing must be frozen. Industrialized nations such as the United States and those in the European Union are still producing huge amounts of OPC to maintain their development and economic growth. Thus, alternative strategies must be taken to lessen the large-scale production of cement-based concretes. Geopolymer (GP) technology is one of the most viable solutions. Presently, GP binders can provide about an 80% reduction in CO_2 emissions compared with OPC. The total CO_2 emissions of GP binders includes the CO_2 emissions of the dissolved solids of the alkaline activators ($Na_2O + SiO_2$) (Miranda, 2005; Criado, 2007, Bakharev, 2005a; Jiminez, 2004). Moreover, the other alternative solutions for reducing CO_2 emissions are the utilization of waste materials from landfill such as fly ash (FA) and slag.

SUSTAINABILITY ISSUES OF CONCRETES

Hydraulic cement, water, and various aggregates such as rock and sand are used to manufacture concretes. Usually, concretes are formulated by means of hefty quantities (about 70%–80%) of aggregates (coarse and fine) (Neville, 2011). The same components in cement are only about 12% (Mehta, 2001). The majority of such ingredients are natural materials obtained via mining or products/by-products manufactured in the industries. An accurate evaluation of the impact of concretes on the environment requires a consideration of their individual components. The collection of coarse and fine aggregates, which involves intensive mining, is also largely responsible for environmental pollution. This is because the extraction of enormous amounts of aggregates from mines, the subsequent processing of raw materials, and transportation all involve the consumption of a significant quantity of energy. Thus, the environment is badly affected (especially forests and riverbeds). It is important to note that about 5%–8% of the total man-made emissions of CO_2 into the environment are related to the manufacturing of OPC, which is regarded as the major binder of concrete, thereby contributing to global warming (Flatt, 2012).

Sand is a naturally occurring granular material comprised of particles of fine alienated rocks and minerals and exists everywhere on earth's landscape in diverse settings. Sand is abundant in ground soil, in streams and flood plains, and in the beds of oceans and rivers (Kondolf, 2008). It is widely utilized in the construction sector worldwide. In the context of Malaysia, the consumption of river sand is rising at a pace far exceeding the limit projected by economic growth rates or the construction industries. The mining of sand implies its removal from the source to the consumer end (Pettijohn, 2012). Irrespective of geographic location, the ever-increasing demand for sand by the construction sector has put immense pressure on the natural sources of sand. Consequently, the removal of sand from various parts of the earth's crust has negatively affected ecological balance and sustainability. Such impacts range from the collapse of riverbanks to damage to habitats, the frequent flooding of plain lands, the destruction of landscapes, the deposit of sediments, the infertility of cultivation land, and the depletion of greenery or vegetation (Kori, 2012). Actually, flood plains and riverbeds are the most profitable resources of gravel and sand.

It is well known that the use of sand in concrete has paramount significance from the standpoint of industrial (construction sector) growth. Earlier studies (Kondolf, 2008; Kori and Mathada, 2012) have revealed that the mining of sand in streams degrades the quality of water and causes the utter destruction or degradation of bed channels as well as riverbanks. Furthermore, sand mining on flood plains to extract fine aggregates also impacts the water table and changes the extent of agricultural land. Sand mining in rivers, channels, in-streams, and flood plains directly influences the functioning of river ecosystems. On top, the wrong mining techniques and related machineries can disturb the ecological balance, where changes in available land are the most common impact on the environment. This reduction of land usage mostly occurs due to the conversion of the unused or innate land into mines/excavations within the earth's surface (Pettijohn, 2012). According to Ismail (2013), the excessive utilization of natural aggregates for the manufacturing of concrete by the construction industries of Malaysia is partially responsible for the destruction of the biodiversity and ecosystems of the nation.

Recently, efforts have been made to reduce the impact on the environment. Industrial concrete production is reduced by preserving natural material resources, lowering the energy required for concrete processing, and improving the concrete's resilience as well as endurance. Previous research has shown that waste materials such as recycled aggregate, coal bottom ash, foundry sand, glass cullet, and slag can be incorporated into concrete as a replacement for sand for sustainability. However, these waste materials have strength and durability issues.

Saifuddin and Purohit (2015) reported the strength properties of SCGPC using manufactured sand. The results show that the contribution of slag helps the mix to attain early and rapid strength development. However, no adverse effects were observed when SCGPC mixes are prepared with manufactured sand. Ushaa (2015) investigated the engineering properties of SCGPC. The results show that all the SCGPC mixes had a satisfactory performance in the fresh state. The blast furnace slag series had good workability compared with the silica fume series. Richardson (2009) prepared a series of concretes using ungraded recycled aggregates and compared their strength with those produced from natural sand. The former (made by recycling) revealed lower compressive strength than the latter (made from natural sand). Meyer (2009) reported that the compressive strength of concretes produced via aggregate recycling ranges within 15%–40% of natural sand-based concretes.

Sagoe (2002) evaluated the permanence and performance features of concretes made from aggregate recycling. It was found that the drying shrinkage and water absorption of concretes made from aggregate recycling were higher compared with natural sand-based ones. This enhancement was attributed to the existence of extra-porous chunks of mortar residue inside the recycled aggregates. Zaharieva (2004) acknowledged that the elevated rate of absorption of recycled aggregates is a major hindrance to the production of concretes. This is because the recycled aggregate-based fresh concretes lose their early workability very quickly despite the use of super-plasticizers. The properties of concrete made using recycled aggregates are mainly ascribed to the attachment of residual mortar within the actual aggregates (Rao, 2007), although the high porosity of the aggregates are unwanted in terms of concrete quality and permanence. Aggarwal and Siddique (2014) observed that concretes with 10%–60% foundry sand and bottom ash in equal proportion would have reduced mechanical strengths. They also reported that the penetration of chloride ions in these concretes was higher than the control. Using recycled fine aggregates together with FA and steel slag, Aanastasiou (2014) prepared several concretes. It was demonstrated that using fine aggregates from demolition and construction waste could increase the porosity and reduce the mechanical strength of concrete. However, by combining these aggregates with steel slag, the lost strength can be recovered. Park (2004) determined the mechanical strength of concretes made of aggregates of waste glass in place of natural sand. The measured slump and compacting factors revealed a reduction. This was due to the angular shape of the grain and an increase in the air content inside innumerable numbers of tiny particles present inside the glass waste. The mechanical properties (e.g., compressive, tensile, and flexural strengths) of these concretes were reduced with increasing concentration of glass waste. The percentage of sand replaced by waste materials was calculated based on experimental results. The optimum percentage of replacement of fine aggregate that will give maximum compressive strength has no specific standard. For this analysis, researchers have considered different percentages of fine aggregate according to the available literature. Table 2.1 summarizes the replacement percentage for different waste materials.

TABLE 2.1
Sand Replacement from Industrial Waste Materials

Author	Industrial Waste	% Replacement
Prabhu et al. (2014)	Waste foundry sand	0, 10, 20, 30, 40, 50
Aggarwal and Siddique (2014)	Waste foundry sand + bottom ash	5+5=10, 15+15=30 20+20=40, 25+25=50, 30+30=60
Devi and Gnanavel (2014)	Steel slag	0, 10, 20, 30, 40, 50
Al-Jabri et al. (2011)	Copper slag	0, 10, 20, 40
Meenakshi and Ilangovan (2011)	ISF slag	0, 10, 20, 30, 40, 50, 60, 70
Tripathi et al. (2013)	GBF slag	0, 10, 20, 30, 40
Shettima et al. (2016)	Iron ore tailings	0, 25, 50, 75, 100
Panda et al. (2013)	FeCr slag	0, 20, 40, 60, 80, 100

GEOPOLYMERS

The following sections include detailed discussions on GPs based on the terminology and the concept of the geopolymerization process.

TERMINOLOGY

Davidovits (1978) was the first to introduce the concept of GPs. It was demonstrated that the reaction between an alkaline solution, silicon (Si), and aluminum (Al) in a geological resource material or in the by-products, including FA and rice husk ash (RHA), can form binders. Later, it was shown that polymerization is the chemical reaction involved in obtaining these geological binders, thereby introducing the word *geopolymer* (Davidovits, 1999).

GPs can exist in three fundamental forms, as follows (Davidovits, 1999):

1. Poly(sialate), represented by the repetitive chemical unit [-Si-O-Al-O-]
2. Poly(sialate-siloxo), represented by the repetitive chemical unit [-Si-O-Al-O-Si-O-]
3. Poly(sialate-disiloxo), represented by the repetitive chemical unit [-Si-O-Al-O-Si-O-Si-O-]

Poly(sialate) as the chemical labeling of GP paste was suggested depending on the ratio of Si to Al or silico-aluminates. The word *sialate* refers to Si-oxo-aluminate. Polysialates represent polymers with chains and rings in the presence of Si^{4+} and Al^{3+} ions in IV-fold taxonomy associated to oxygen that extends from disordered (amorphous) to ordered (semi-crystalline) structures. Figure 2.1 illustrates the types of GP.

Figure 2.2 presents a conceptual model for the geopolymerization process where solid alumino-silicate is transformed into synthetic alkaline alumino-silicate. Aluminate and silicate species are produced due to the dissolution of the solid alumino-silicate via the hydrolysis of alkali solution. The species are activated by the high-solution pH. First, the super-saturation of alumino-silicate gives rise to a gel that further forms a large network of oligomers via condensation. Water is released in the formation of the GP that is further excluded from the GP matrix during the curing and drying processes (Davidovits, 1999). Consequently, discontinuous

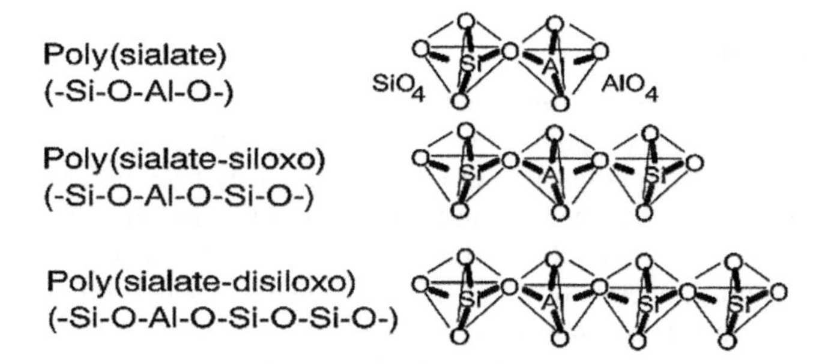

FIGURE 2.1 Structural units and terminology of GP (Davidovits, 1994).

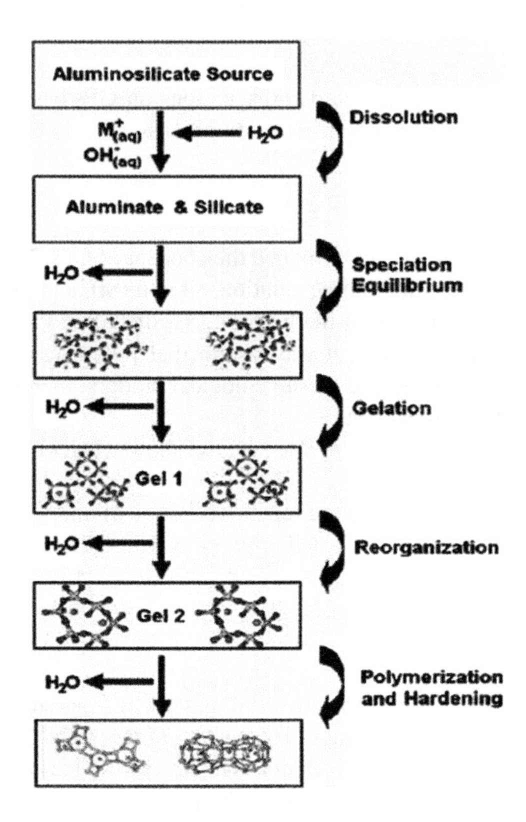

FIGURE 2.2 Schematic representation showing the different steps of the geopolymerization process (Fernández-Jimenez, 2006).

nanopores are left over within the GP matrix, which in turn enhances the performance of the GP. In short, the presence of water in GP mixture has no role in the chemical reaction. Water merely renders the mixture's workability during treatment. Conversely, water participates heavily in the chemical reaction of OPC mixture during the hydration process.

Geopolymer Constituents

Source Materials

Any material containing the disordered (amorphous) form of Si and Al is a prospective source material for GP production. The prospect of diverse minerals and industrial by-products in the production of GP has previously been investigated. These include metakaolin or calcined kaolin (Davidovits, 1999; Barbosa, 2000; Hardjito, 2005), low-calcium ASTM class-F FA (Palomo, 1999; Swanepoel and Strydom, 2002), the natural minerals of Al-Si (Xu and van Deventer, 2000), a mixture of non-calcined and calcined minerals (Xu and van Deventer, 2000), a blend of FA and metakaolin (Swanepoel and Strydom, 2002; van Jaarsveld et al., 2003), and a mixture of ground granulated blast furnace slag (GGBFS) and metakaolin. GGBFS is an industrial by-product of steel and pig iron, which is obtained from the blast furnace via the reduction process of iron ore into iron. In this process, the trapped molten slag from the blast furnace is quenched very quickly in the presence of water and then dried as well as ground into fine powdered form. Through such fast quenching, the slag is fragmented into a glassy granular structure to achieve cementitious properties (Dinakar, 2013).

Generally, the chemical composition of slag involves simple oxides that are calculated from X-ray fluorescence data-based elemental analysis. The quenching rates of slag during the processing of steel and pig iron greatly determine the slag's mineral structures. In the case of steel slag, the rate of cooling is adequately low for obtaining crystalline compounds. The major compounds in steel slag, as listed in Table 2.2, are dicalcium silicate (Ca_2SiO_4), dicalcium ferrite ($2CaO{\cdot}Fe_2O_3$), calcium aluminate (Al_2CaO_4), calcium-magnesium iron oxide, and some free lime as well as free magnesium (Mg). The ratios of these compounds are decided by the steel manufacturing practice and the rate of cooling (Teng, 2013).

Molarity of Alkali Activator

Chemical activator or alkali activator solution plays a major role in the initiation of the geopolymerization process. Generally, a strong alkaline medium is necessary to increase the surface hydrolysis of the aluminum and silicate particles present in the raw material, while the concentration of the chemical activator has a pronounced effect on the mechanical properties of the GPs (Hum, 2009). On the other hand, the dissolution of silicate (Si) and aluminate (Al) species during the synthesis of GPs is very much dependent on the concentration of NaOH, where the amount of Si and Al leaching is mostly governed by the NaOH concentration and the leaching time (Panias, 2007). Gorhan and Kurklu (2013) reported the influence of NaOH solution on the compressive strength over 7 days of FA GP mortars subjected to different NaOH concentrations. Three different concentrations of NaOH (3, 6, and 9 M) were used throughout the laboratory work. Based on the results acquired, the NaOH concentration that produced the highest 7-day compressive strength of 22.0 MPa was 6 M. When the NaOH concentration was too low (3 M), it was not sufficient to stimulate a strong enough chemical reaction, while an excessively high concentration of NaOH

TABLE 2.2
Typical Chemical Compositions of Slag

Constituents	Composition (%)
Calcium oxide (CaO)	32–45
Silicon dioxide (SiO_2)	32–42
Iron oxide (Fe_2O_3)	0.1–0.5
Manganese oxide (MnO)	0.2–1.0
Magnesium oxide (MgO)	5–15
Aluminum oxide (Al_2O_3)	7–16

Source: Das (2007).

(9 M) resulted in the premature coagulation of silica, which in both cases culminated in lower-strength mortars. Somna (2011) studied the compressive strength of ground FA cured at ambient temperature by varying the NaOH concentration from 4.5 to 16.5 M. Results showed that by increasing NaOH concentrations from 4.5 to 9.5 M, a significant increase in the compressive strength of the paste samples is observed, while the variation of NaOH concentrations from 9.5 to 14 M also increases the compressive strength of the paste samples but increases the viscosity of the GP, leading to low workability. The increase in compressive strength with the increasing NaOH concentrations is mainly due to the higher degree of silica and alumina leaching. The compressive strength of ground FA-hardened pastes starts to decline at NaOH concentrations of 16.5 M. This decrease in compressive strength is mainly attributed to excess hydroxide ions, which cause the precipitation of alumino-silicate gel at very early ages, thus resulting in the formation of lower-strength GPs.

The purity of chemical activators such as sodium/potassium hydroxide and soluble silicates in the mix design of GPs has a significant effect on the properties of GPs. According to Part (2015), the nature and purity concentration of the alkali activator is the most dominant parameter in the alkali activation process. The authors utilized five different types of alkali activators ($Ca(OH)_2$, NaOH, NaOH + Na_2CO_3, KOH, and Na_2SiO_3) with various purity concentrations to fabricate FA-based GP mortars. Based on the compressive strength results, the alkali activator that possesses the highest activation potential and highest compressive strength was Na_2SiO_3, followed by $Ca(OH)_2$, NaOH, NaOH + Na_2CO_3, and KOH. The lower activation potential of KOH compared with NaOH was due to the difference in ionic diameter between sodium and potassium. Fadhil Nuruddin (2011) reported the use of Na_2SiO_3 (grade A53) with 55.52% of water, 29.75% of SiO_2, and 14.73% of Na_2O to create an activator solution, with pellets of NaOH (99% purity) to avoid the effects of unknown contaminants in the mixing water.

Super-Plasticizers

To manufacture long-lasting concretes, it is important to design concrete mixes with three vital features. The concrete mixes must have high workability to guarantee their placement and compaction in highly congested reinforced regions with the highest compaction and density. Concrete mixes must have elevated early-age strength development in terms of being free of both early-age stripping and cracking. Lastly, such concretes must have low ratios of water binder, which is a major contributor to durability. Actually, super-plasticizers are ingredients that can accomplish all the aforementioned necessary characteristics of concrete. Thus, super-plasticizers are often considered an indispensable component of concrete for great durability and very high workability (Kong, 2016).

Nuruddin (2015) acknowledged that the premixing of alkaline solution, super-plasticizer, and extra water prior to their addition into dry concrete mixes can effectively improve the workability and strength of the SCGPC. According to Memon (2012), the inclusion of extra amounts of both a super-plasticizer and water could remarkably improve the workability and compressive strength of SCGPC. Conversely, the incorporation of more than 15% water could cause in-bleeding and segregation, which in turn reduces the compressive strength of the concrete.

Meanwhile, the addition of water beyond 12% by mass of FA could drastically reduce the compressive strength of SCGPC.

Mixture Proportions

About 75%–80% by mass of concrete consists of aggregates. GPCs are manufactured using similar tools to normal concretes. The following parameters affect the workability and mechanical strength of concretes (Hardjito, 2005). Increases in sodium hydroxide contents enhance the concrete's mechanical strength. Increases in the ratio of sodium silicate to NaOH solution augments the concrete's mechanical strength. The addition of extra water into the mixture raises the concrete's slump value. The addition of extra super-plasticizer improves the concrete's workability.

Fresh Geopolymers

Presently, information on the behavior of fresh GPs is very limited. Metakaolin-based fresh geopolymer mortars (GPMs) are found to be very rigid, dry while mixing, highly viscous, and cohesive in nature (Teixeira-Pinto et al., 2002). It is suggested that rather than gravity-type mixers, forced mixers must be employed during the mixing of GP materials. Furthermore, increases in the mixing time raise the temperature of the fresh GP and reduce the workability. To improve workability, the use of admixtures is suggested, which may diminish viscosity and cohesion. For metakaolin- and GBFS-based fresh GP paste, the setting times at room and elevated temperatures (in an oven) are measured. The initial setting time of GPs cured at 60°C was observed to be very short, ranging between 15 and 45 minutes. To achieve good results in concrete placement, the retention time of the rheological properties of SCGPC is greatly significant (Hardjito, 2005). This retention time can be adjusted by selecting appropriate types of super-plasticizers or combinations of retarding admixtures. The different effects of diverse admixtures on the open time of SCGPC can be utilized depending on the kind of cement, the transport time, and the placement duration of the SCGPC (EFNARC, 2002).

Factors Affecting Geopolymerization and Geopolymer Color

The fresh and hardened properties of GP paste are largely decided by the characteristics of the constituent materials and their chemical ratios, which are similar to that of PC-based concrete. Information on the diverse properties of GP is necessary to create a universal guidance, which may promote explorations on GPs depending on the justification and confirmation of the established knowledge (Hardjito, 2004). The color of GPs as well as the geopolymerization process are extremely susceptible to the constituent raw materials, such as the distribution of particles in terms of size and shape, the degree of crystallization, the types of alkali activators used (NaOH, KOH, ratio of sodium to potassium silicates, etc.), the various conditions of curing (temperature, degree of moisture, condition of opening or healing, time of curing, etc.). Highly durable GP cement with diverse mechanical, microstructural, and thermal properties can be manufactured based on altered constituent raw materials, alkali activator solutions, and conditions for curing (van Jaarsveld, 2003). The most important thing is the presence of a large fraction of alumina and silica in the raw materials, non-metallic inorganic minerals, and industrial waste, which are the major active ingredient in aluminum silicate, to form the SCGPC. To produce a good quality of geopolymer concrete (GPC), various raw materials can be employed, including FA, red mud, metakaolin, natural pozzolan, blast steel slag, rice HA, and so on. Likewise, several choices are adopted for alkali activators. Alkali metal hydroxide (sodium hydroxide), carbonate, sulfate, phosphate, and fluoride (a few studies) can be used as activators.

Higher NaOH dosages can result in better workability; higher 1-, 7-, and 8-day strengths; and shorter demolding time. Excessive NaOH concentration adversely affects the strength, wherein the optimal NaOH content depends on other mixture constituents. The compressive strength of GPC increases with the increase of Na_2SiO_3 to NaOH liquid proportion by mass. This improvement in the compressive strength can be ascribed to the increased Si to Al ratio due to the availability of extra Na_2SiO_3. The extra water lingers exterior to the GP network and acts as a lubricating agent (Davidovitts, 2005).

The microscopic mechanism of the polymerization process is not yet fully understood. Nevertheless, it is argued that incorporation of water allows alterations in workability without becoming a part of the resultant GP structure. However, curing plays an important role in the geopolymerization process.

Hardjito (2004) investigated the compressive strength of FA-based GP cement at curing temperatures in the range of 30°C to 90°C. The compressive strength was increased with the increase of curing temperature. Furthermore, a longer heat curing time was found to improve the degree of geopolymerization and thereby enhance the compressive strength of the GPC and cause alterations in the color. The curing method plays a very significant role on the overall properties of GPs. The natural curing method with only a plastic film cover is better than the wet cloth-covered natural curing method and the water curing method (Bakharev, 2005b). The curing method of heating without sealing will lead to the quicker evaporation of water and color changes and then cause macro-cracks on the surface of the samples (Yongde and Yao, 2000).

SELF-COMPACTING CONCRETES

For many years after 1983, the issue of the permanence of concrete structures remained a focused research topic in Japan (Kosmatka, 2002). The production of long-lasting concretes needs sufficient compaction and skilled personnel. The designs of contemporary reinforced concrete structures became more advanced, and the shape design became complex with the usual heavy reinforcement. Moreover, the steady reduction in the number of skilled hands in the construction industries of Japan has witnessed a parallel decline in construction work quality. The implementation of self-compacting concretes (SCCs) may be a feasible solution to achieving long-lasting concrete structures, irrespective of the construction work quality. Such concrete structures can be compacted into every corner through their own weight and in the absence of any vibrating compaction (Łaźniewska, 2014). Kamada et al. (1986) proposed the need for SCCs (Ozawa, 1989; Maekawa et al., 1993) at the University of Tokyo, developed some prototype SCCs, and determined their workability. These SCCs were prepared using the available materials in the marketplace, which performed suitably concerning the heat of hydration, drying and hardening shrinkage, post-hardened denseness, and other features (Kosmatka, 2002). Having SCC as the mixture fluid is greatly advantageous for placement under complicated conditions with congested reinforcement and without vibration.

Theoretically, self-compacting or self-consolidating concretes are characterized by the following attributes: they maintain a fluidity that allows self-compaction without external energy, they maintain homogeneity throughout and after the process of placement, and they maintain an easy flow during reinforcement (Dinakar, 2013). Currently, SCCs are used in the precast industries as well as for some commercial applications. Moreover, the comparatively high cost of materials is hindering the widespread use of these concretes in the construction industries, especially for commercial and residential building sectors (Lazniewska, 2015). SCCs are far more expensive than traditional concretes because of the great demand for cementitious materials and chemical admixtures such as high-range water-reducing admixtures (HRWRA) and viscosity-enhancing admixtures (VEA). The cementation materials in SCCs for filling very restricted areas and repairing applications usually range from 450 to 525 kg/m^3. These applications need low volumes of aggregates for the flow between restricted spaces to be free from obstruction and to ensure the formwork is filled without consolidation. The inclusion of finely ground powdered materials in high volumes is essential to improve the cohesion where high paste volumes are needed for the successful casting of SCCs (Kosmatka, 2002).

Advantages of Self-Compacting Concretes

As mentioned previously, SCCs do not necessitate any vibrations to accomplish complete compaction. This offers the following notable advantages over other concretes (Kosmatka, 2011): improved concrete quality and decreased onsite repair, a more rapid construction process, overall cost lowering, improved health and safety in the absence of any vibrators, a significant drop in environmental noise loading on and around sites, the feasibility of using dusts and waste products that are expensive to dispose of.

Limits on Self-Compacting Concrete Material Proportions

Table 2.3 summarizes the limits on the material ratios of SCC (Kosmatka, 2002). The limits of the cementitious materials for high fines are 450–600 kg/m^3, while the combination is between 385–450 kg/m^3. The water–cement ratio for high fines is 0.28–0.45. The fine aggregate percentage for high-fines materials is between 35%–45% and 40% for combination materials.

European Guidelines for Self-Compacting Concrete (EFNARC)

The following important factors must be taken into account relating to the relative ratios of the main components of the blend by volume rather than by mass during the mix design: The water-to-powder proportion by volume must be within 0.80–1.10. The overall powder content must range between 160 to 240 L/m^3 or 400 to 600 kg/m^3.

TABLE 2.3
Limits on SCC Material Proportions

Materials	High Fines	Combination
Cementations (kg/m^3)	450–600	385–450
Water/cementations material	0.28–0.45	0.28–0.45
Fine aggregate/mortar (%)	35–45	40
Fine aggregate/total aggregate (%)	50–58	–
Coarse aggregate/total mix (%)	28–48	28–48

Source: Kosmatka et al. (2002).

Coarse aggregate content usually varies from 28% to 35% by volume of the blend. The water-to-cement ratio should be considered depending on EN 206 recommendations. Usually, water content must not exceed 200 L/m^3. The sand content must balance the volume of the other components. The requirements and guiding principles of the European Federation of National Associations Representing for Concrete (EFNARC) for workability tests of SCC (slump, L-box, V-box, and T50) must be followed. Commonly, it is desirable to have a conservative design to guarantee that the concrete can maintain its definite fresh characteristics even though changes are expected in the quality of the raw material. Alterations in the moisture content of the aggregates must be considered during the design of the mix. Usually, admixtures that can modify the viscosity are preferred to compensate for the changeability of the aggregate's sand grades and moisture contents in the mix.

Requisites in the Mix Design

High Volume of Paste

It is important to note that the SCC's spreading and filling capacity is limited by the friction among the aggregates. For this reason, the SCC must enclose high volumes of paste comprising cement, additives, and enough water and air (typically in the range of 330 to 400 L/m^3) to maintain the aggregate's isolation (EFNARC).

High Volume of Fine Particles (Size below 80 μm)

To ensure the adequate workability of the SCC by restraining the possibility of segregation or bleeding, it must enclose an outsized number of fine particles. However, to circumvent the generation of excess heat in OPC, it is partly replaced by mineral admixtures such as limestone filler or FA or GGBFS. The added fillers' characters and quantities are selected to fulfill the requirements of strength and workability (EFNARC).

Plasticizers of High Dosage

To attain the fluidity of the SCC, super-plasticizers are incorporated. Nonetheless, a high dose close to the saturated amount can enhance the limitations related to concrete segregation (EFNARC).

Low Volumes of Coarse Aggregate

SCC can be produced using natural aggregates of rounded, semi-crushed, or crushed types. However, the performance of SCC in congested areas is decided by the coarse aggregate's features, where volume limitation is the key factor. Actually, the maximum size of the aggregate (Dmax) must be in the range of 10 to 20 mm (EFNARC).

GARNET

Garnet History

So far, no detailed literature review has been performed on spent garnet material that shows its potential for concrete applications. In the context of Malaysia, spent garnet is becoming a significant issue in terms of recycling this new waste material and further reusing it as a useful replacement for sand in concrete. Statistics have revealed that the amount of garnet imported from Australia to Malaysia in 2013 was 2000 metric tons. This amount was imported by Malaysia Marine and Heavy Engineering (MMHE). However, heavy mineral sands are a significant class of ore deposit as they are a resource of rare earth elements and industrial minerals, including diamond, garnet, sapphire, and other valuable gemstones or metals (Roskill, 2000).

Generally, in beach environments, placer deposits are formed by concentration because of the specific gravity of the mineral grains. Although most of the heavy minerals (other than gold placers) are present inside stream-beds, the majority of them are relatively smaller and lower grade. The cutoff grades of heavy minerals as the total heavy mineral (THM) concentrate from the raw sand in most ore deposits of this type will be about 1%. Usually, THM concentrate is in the range of 1% to 50% for zircon, 10% to 60% for ilmenite, 5% to 25% for rutile, and 1% to 10% for leucoxene. The other bulk of the THM content includes magnetite, chromite, and garnet, which are considered trash minerals. Sand minerals from beaches have lately been shown to have deposits of minerals that originated in the southern hemisphere because of continental drifts. Because of the movement among diverse plates and frequent weathering (heating and cooling), rocks develop fractures that allow water to diffuse. Thus, parent rocks face erosion and weathering and are then transported along with sea water before becoming sediment in an appropriate basin. Continuous attrition mediated by air and water occurs, which escalates with time. These resources of heavy mineral sands are in the erosion areas of rivers, where the eroded minerals are dumped into the ocean. Subsequently, such sediments are involved in littoral or longshore drift. Often, the heavier rocks are eroded directly by the ocean waves and wrecked on the beaches, and the lighter ones are blown by the wind (Harris, 2000). The compositions of the precious minerals are determined by the source rocks. Generally, zircon, rutile, monazite, and some ilmenites originate from granite. Ultra-mafic and mafic rocks of garnet are the main resource of ilmenite, including kimberlite or basalt. Garnet is resourced from the metamorphic rocks including amphibolite schists. Precious metals are resourced from deposits enclosed inside metamorphic rocks (Krishnan, 2016). Ali et al. (2001) reported that India is gifted with precious resources of beach sand minerals where the coastal line of Tamil Nadu, Odisha, and Andhra Pradesh are enriched with considerable amounts of deposits of various minerals.

These precious minerals include ilmenite ($FeO.TiO_2$), rutile (TiO_2), monazite ([Ce,La,Y,Th]PO_4), garnet, sillimanite ($Al_2O_3.SiO_2$), and zircon ($ZrO_2.SiO_2$). These resources of placer minerals in India were estimated (in millions of tons, MT) as follows: 348 MT of ilmenite, 107 MT of garnet, 21 MT of zircon, 18 MT of rutile, and 130 MT of sillimanite. Rajamanickam (2004) acknowledged that the estimated total global reserve of placer minerals is about 1775 MT, where India alone has 278 MT of ilmenite, 13 MT of rutile, 18 MT of zircon, 7 MT of monazite, 86 MT of garnet, and 84 MT of sillimanite.

For centuries, garnets have been employed as gemstones. In recent times, it has been discovered that the occurrence of angular fractures, relatively high hardness and specific gravity, and the recycling ability of garnet make this material greatly advantageous for several industrial applications. Garnet can be exploited for a broad range of purposes. Usually, garnet is a generic name coined for an assemblage of complex silicate minerals with isometric crystal structures and comparable chemical compositions. The general chemical formula for garnet minerals is denoted by $A_3B_2(SiO_4)_3$, where A can be calcium, magnesium, ferrous iron, or manganese and B can be aluminum, chromium, ferric iron, or rarely, titanium. The six most commonly occurring garnet minerals are classified into three groups: (1) aluminum garnets, (2) iron garnets, and (3) chromium garnets. The most frequent minerals of the aluminum garnet set are almandine or almandite, grossularite, pyrope, and spessartite. The most common iron garnet mineral is andradite, and uvarovite is the most widespread chromium garnet (Krishnan, 2016).

On earth, garnet appears in numerous rock types, mainly gneisses and schists. Other sources of garnet minerals are contact metamorphic rocks, crystalline limestones, pegmatites, and serpentinites. Alluvial garnet is related to the sand of heavy minerals and deposits of gravels worldwide. Our planet is plentiful with diverse types of garnet. Comparatively few commercially practicable garnet deposits have been discovered and exploited so far. The complex mineralogy of garnet shapes the broad array of applications from filtration media to water jet abrasives. The garnet industry of the United States has hardly any major producers. The market values of potentially industrial garnet are decided by the reserves' grades, sizes, mined types, qualities, and the closeness of the mineral deposits to the developed infrastructure and users, as well as the costs of milling (Ali, 2001).

The price tag of garnet within U.S. industries is exceptionally competitive, where the suppliers are required to render a high standard of customer service. The majority of industrial-grade garnets that are mined in the United States are almandine and pyrope (magnesium-aluminum silicate). Several andradites (calcium-iron silicates) are also mined domestically for industrial purposes. The United States is the principal consumer of industrial garnet in the world and consumes over 25% of industrial garnet globally (Harris, 2000). In 2002 alone, it consumed about 54,200 tons of industrial garnet, and it was 9% higher in 2002 than in 2001. Still, the growth rate of the water jet-cutting sector is slightly higher compared with the abrasive blasting industry. The foremost uses of garnet in the United States and their projected market distribution in 2002 were as follows: abrasive blasting media 35%, water jet cutting 30%, water filtration 15%, abrasive powders 10%, and another end uses 10% (Gorrill, 2003).

The main consumers in the domestic industrial sectors are aircraft and motor vehicles, ceramics and glasses, electronics, filtration plants, petroleum, shipbuilding, and wood furniture finishing. Most industrial garnet is used in the form of loose grain abrasive due to its hardness (6.0 to 7.5 on the Mohs scale). Lower-grade industrial garnet is usually employed as a filtration medium in water purification systems due to its chemical inertness and resistance to degradation. Principally, high-quality and good-value garnet grains have been used over centuries for diverse applications such as the grinding of optical lenses and glass plates. Industrial diamond and fused aluminum oxides are contenders for such applications. Currently, industrial garnet powders are used to make semiconductors and metals with high-quality and scratch-free lapping. Further appliances are related to the production of abrasive coatings, hydro-cutting, wood finishing, leather smoothening, and the hardening of rubbers, felts, and plastics. Due to its non-toxic nature, garnet is gradually replacing silica sand in the blast-cleaning market. Garnet is free from health hazards related to inhalation such as airborne crystalline silica dust inhalation. Presently, silica sand and mineral slag are the predominant media used for blasting purposes (Ryan, 2000).

The petroleum sector of the United States is one of the largest garnet consumers, employing garnet for cleaning drilling pipes and well casings. The price increase of crude oil in 2002 meant that extra equipment maintenance was required in the petroleum sector, where many industrial garnets were used. Shipbuilders and aluminum aircraft manufacturers use large amounts of garnet for blast cleaning and metal surface finishing. Other applications include the cleaning and conditioning of aluminum and various soft and hard metals, and structural steel fabrication. The relatively heavier nature and good chemical stability of garnets make them suitable for water treatment and filtration applications. To get better water quality in a more reliable way, mixed-medium water filtration system are developed using a blend of sand, anthracite, and garnet. This new mixture is gradually replacing older filtration systems. In fact, as a filtration medium, garnet is competing with silica sand, magnetite, ilmenite, and plastics (Olson, 2001).

Garnet entrained in high-pressure water streams are also used for the cutting of various materials. Generally, garnet powders are used for polishing glass and ceramic and painting anti-slip and anti-skid surfaces. In the abrasive coating marketplace, garnet ranges from inexpensive quartz sand or staurolite to expensive abrasives, including silicon carbide and fused alumina. For most abrasive coating purposes, garnet is more preferred than quartz sand because of its friable nature and lower hardness. However, for metalworking applications that involve extensive metal removal, garnet is unable to compete with currently produced abrasives. The broad price range of industrial garnet depends on the specific use, quality, amount, resource, and type. The average selling price of crude concentrates per ton in 2002 ranged from approximately \$53 to \$220, with an overall domestic average of \$117 per ton. Conversely, the average cost per ton of refined garnet sold in this year varied from \$61 to \$331, with an overall domestic average of \$165 per ton, which was about 5% lower than in 2001 (Gorrill, 2003).

In 2002, the total estimated global production of industrial garnet was 440,000 tons, and China, Australia, India, and the United States were the major producers. The United States alone produced about 9% of the total industrial garnet produced worldwide. Production in both Australia and India was greater than the United States. Currently, Russia and Turkey are mining garnet mainly to meet their domestic market demand. Besides, some countries with garnet resources such as Chile, Canada, Czech Republic, South Africa, Pakistan, Spain, Ukraine, and Thailand have established small mining facilities to meet their domestic demand. Industries in Australia have been steadily enhancing their garnet manufacturing and export. Meanwhile, China and India have augmented garnet production and become vital suppliers of garnet for other nations (Jeffrey, 2006).

Physical and Chemical Characteristics of Garnet

Varieties of garnet minerals exist, each having its characteristic chemical composition. The predominant minerals in the garnet group are almandine, pyrope, spessartine, andradite, grossular, and uvarovite. All these minerals possess vitreous luster, varying diaphaneity from transparent to translucent, brittle tenacity, and a lack of cleavage. They occur in the form of individual crystals, stream-worn pebbles, granular aggregates, and massive clusters (Evans, 2006). The chemical formula, hardness, specific gravity, and colors of these garnets are summarized in Table 2.4.

Among the major garnet minerals, the wide range of solid solution series decides the various physical properties of garnet. For instance, calcium garnets usually possess low specific gravity and low hardness and are green in color. Conversely, garnets of iron and manganese possess high specific gravity and high hardness and are red in color (Evans et al., 2006). In this study, almandine garnet was selected due to its abundance in Malaysia. Table 2.5 lists some chemical properties of diverse garnets that occur worldwide.

Utilizations of Garnet

Garnets have been utilized as gemstones for thousands of years, but in the past few decades garnets have been exploited for several industrial applications. Presently,

TABLE 2.4
Important Physical and Chemical Properties of Some Garnet Minerals

		Garnet Minerals		
Mineral	**Composition**	**Specific Gravity**	**Hardness**	**Color**
Almandine	$Fe_3Al_2(SiO_4)_3$	3.0	7.0–7.5	Red, brown
Pyrope	$Mg_3Al_2(SiO_4)_3$	3.56	7.0–7.5	Red to purple
Spessartine	$Mn_3Al_2(SiO_4)_3$	4.18	6.5–7.5	Orange to red to brown
Andradite	$Ca_3Fe_2(SiO_4)_3$	3.90	6.5–7.0	Green, yellow, black
Grossular	$Ca_3Al_2(SiO_4)_3$	3.57	6.5–7.5	Green, yellow, red, pink, clear
Uvarovite	$Ca_3Cr_2(SiO_4)_3$	3.85	6.5–7.0	Green

Source: Evans (2006).

TABLE 2.5
Chemical Properties of Various Garnets

	Source						
Chemical Formula	**Sabah, Malaysia**	**Beni Bouchera, Morocco**	**Alaska, United States**	**Lers, France**	**Ronda, Spain**	**Hawaii, United States**	**Indian Beach**
Aluminum as Al_2O_3	15.8	13.80	17.78	14.76	16.16	10.89	19.9
Silica as SiO_2	44.4	45.35	40.43	46.64	47.63	47.73	36.15
Calcium as CaO	15.7	12.76	15.31	14.55	14.15	12.31	3.41
Magnesium as MgO	12.2	13.45	10.65	13.85	9.88	16.65	7.22
Iron as Fe_2O_3	2.0	1.8	3.2	1.26	0.94	2.98	32.86
Loss of ignition	–	–	–	–	–	–	–

Source: Schmolke (2006).

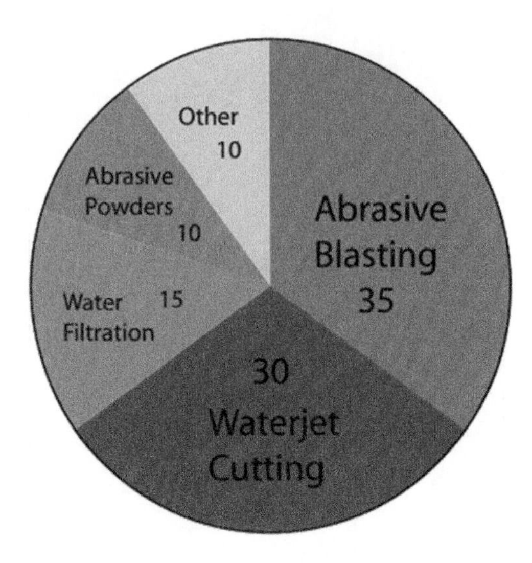

FIGURE 2.3 Different uses of garnet in the United States (Olson, 2005).

garnet is emerging as a significant industrial mineral for assorted applications. Figure 2.3 demonstrates some recent industrial uses of garnet in the United States. It is worth noting that garnet is often used as an indicator mineral during mineral exploration and geological evaluations.

INFLUENCE OF WASTE MATERIALS ON THE ENGINEERING PROPERTIES OF CONCRETE

Fresh Properties

The majority of the available literature on concrete containing waste materials as fine aggregates shows their lower workability than concretes with natural fine aggregates such as sand. It has also been observed that workability is affected at certain levels of replacement. These may be attributed to the shape, size, and surface characteristics as well as particle size distribution, which could modify the behavior of fresh concretes (Evangelista, 2007). Concrete workability is directly correlated to the fineness of the materials used for concrete production (Prabhu, 2014). According to EFNARC (2002), the rank of fluidity of the SCC is decided mainly by the super-plasticizer dosages. Overdosing often causes segregation and blockage. The properties of fresh SCC require careful control, and two types of tests are performed to ensure this aspect. Based on workability test results, it has been reported (Thampi et al., 2014) that the sodium hydroxide (pellet) molarity concentration must be kept at 12 M to achieve good workability.

Zhao (2015) studied the workability properties of SCGPC using tests including J-ring, V-funnel, L-box, and T50. It was shown that the workability properties of Geopolymer concrete containing 20% FA replacement with cement and 5% super-plasticizer could satisfy the EFNARC guidelines. The influence of curing temperature and time on the properties of GPC was studied. Furthermore, the influence of excess water on the workability of GPC cubes was examined. FA and GGBFS were utilized as binders, which were blended with an alkaline solution to produce GP paste rather than cement paste to bind the aggregates. Experimental results revealed that an increase in the ratio of water content to GP solids could increase the workability of GPC (Nagral, 2014).

Zhao et al. (2014) examined the feasibility of using iron ore tailings in place of natural fine aggregate to produce ultra-high-performance concretes (UHPCs). It was observed that by replacing 100% of the natural fine aggregate with tailings could considerably reduce the workability and compressive strength of the concrete. This degradation of the properties was ascribed to the higher specific surface area of tailings than natural sand, which could absorb higher amounts of water. Goyal et al. (2015) reported that an increased percentage of iron dust in the concrete mix could reduce the workability where the use of super-plasticizers was recommended. Kang et al. (2011) reported that with concrete grades above C40, the concrete slump could reduce swiftly, leading to a lowering of the concrete's workability. The inclusion of iron tailings could lower the concrete's slump further than that with only river sand because of the large number of finer particles and the larger specific surface area (Zhang, 2013).

Sreenivasulu (2016) studied SCGPC-based class-F FA and GGBFS with 3% super-plasticizer and 100% manufactured sand as fine aggregates. All mixes were prepared with a constant ratio (by mass) of water to GP solids of 0.4 and a fixed total binder content of 450 kg/m^3. Results revealed that an increase in the NaOH concentration from 8 to 12 M could enhance the viscosity and degrade the fresh properties of SCGPC mixes. The impact of alkali-activated solution (AAS) content on the GPC mixture was reported by Parthiban (2013), where gelenium B233 (2%) was used to enhance the workability of GPC. The workability of GPC was found to enhance with an increase in the slag content.

Compressive Strength

In most concrete material categorizations, the compressive strength of hardened concrete is regarded as a significant property. Nuruddin (2011) acknowledged that a comprehensive strength as much as 54 MPa can be achieved with a curing time duration of 48 hours at a temperature of 70°C. The influence of NaOH molarity on the compressive strength of SCGPC was reported by Memon (2012). A NaOH content variation was found to influence the compressive strength of the SCGPC monotonously, which was increased as the NaOH content in the aqueous phase was increased from 8 to 12 M. For a super-plasticizer content up to 6% and NaOH of 12 M, the value of compressive strength at 7 days was adequate. The strength was as much as 70% at 28 days due to the inclusion of slag (Saifuddin, 2014).

Experiments were performed to determine the effects of alkaline solution variation on the mechanical properties of GPC. It was shown that the compressive strength can be enhanced by reducing the ratios of water to binder and aggregate to binder (Sanni, 2013). The performance of SCGPCs comprised of various mineral mixtures was studied by Ushaa (2015). It was found that for slag replacement of 30% with cement, the compressive strength was enhanced to 38 MPa at 28 days of curing.

Siddique et al. (2009) reported that there was a marginal increment in the compressive strength of waste foundry sand concrete mixture via the replacement of natural sand by 10%, 20%, and 30%. This improvement was ascribed to the fineness of the foundry sand and silica content, which compacted the concrete matrix. According to Aggarwal and Siddique (2014), the use of foundry sand and bottom coal ash at 10%–60% fine aggregate replacement could decrease the compressive strength of the SCGPC compared with the control specimen. However, they showed that the compressive strength of concrete with 50% foundry sand and bottom ash was useful for structural applications. A series of concretes was prepared by blending fine aggregates such as crushed fine stone (CFS), fine recycled aggregate (FRA), and furnace bottom ash (FBA) to study the mechanical properties (Kou and Poon, 2009). At all the ages, the compressive strength of concrete was observed to reduce with increased FBA and FRA contents. This observation was attributed to the elevated content of initial free water in the mixes, which created bleeding and weak interfacial bonding among the aggregates and the cement pastes. Memnon (2012) reported the influence of super-plasticizers and additional water content on the workability and compressive strength of SCGPCs. It was demonstrated that the addition of super-plasticizers not only enhanced the workability properties of fresh concrete but also improved the compressive strength of hardened concrete. In fact, the compressive strength of SCGPC was reduced considerably with excess water content more than 12% by mass of FA.

Flexural and Splitting Tensile Strengths

Splitting tensile strength is one of the most significant mechanical properties in the characterization of the various aggregate bondings in concrete. The structure of concrete is usually influenced by tensile cracking and is an important issue, particularly for concrete slabs in airport and highway implementation (Neville, 2011). Flexural strength estimates the load of the concrete members during cracking. Contrary to popular knowledge on compressive strength, relatively less information is available on the impact of waste materials on flexural and tensile strengths on SCGPC. Ariffin (2015) showed that that inclusion of ceramic aggregate in place of river sand could positively influence the compressive, splitting tensile, and flexural strength of GP concrete at all curing ages. Strength is gained in the samples containing ceramic aggregates as river sand replacement faster than in control samples. In all tests of compressive, splitting tensile, and flexural strength, 100% ceramic aggregate as river sand replacement displayed better results than control specimens.

Ushaa (2015) studied the flexural properties of SCGPC beams by partially replacing FA with GGBFS as well as river sand with M-sand subjected to two-point loading. The crack patterns in the SCGPC beams were identical to those reported with reinforced OPC beams. It is worth mentioning that all fabricated beams failed in flexural tests and showed ductile behavior together with rupturing of the concrete in the compression region. In another study, Sanni (2013) revealed that by increasing the alkaline proportion, the splitting tensile strength was enhanced for any grade of GPC. The best dose of alkali solution was chosen as 2.5 M. At this proportion, the GPC can attain the highest compression and tensile strength, irrespective of grade.

Ugama and Ejeh (2014) reported that for mortar containing 20% iron ore tailings, the flexural and tensile strength was higher compared with the control specimen. Moreover, the strength of the mortar that contained 40%–100% iron ore tailings was lower than the control specimen.

Static Elastic Modulus

The static elastic moduli of concrete are significant mechanical characteristics that determine the relation between stress (force) and the corresponding strain (deformation) of concrete (Neville and Brooks, 2010). In structural analyses of reinforced concretes, the elastic moduli are used to predict moments, stresses, and deflection. For the assessment of a concrete's stiffness and deflection, information about its elastic properties are a prerequisite for the designers. However, there is little or no information in the literature about the elastic moduli of waste materials incorporated into concretes. The elastic moduli of concrete are directly correlated to compressive strength, aggregate types and amounts, and unit weight (ACI 318-02, 1995). Adjustments in the mix proportion, particularly the sand-to-combined-aggregate ratio, are carried out to achieve the desired slump in the fresh state and usually influence the modulus of elastic (ACI 318-02, 1995).

DURABILITY OF CONCRETE

Many studies have been carried out to ascertain and quantify problems with the poor stability and endurance of concrete. Over the years, efforts have been made toward the utilization of diverse waste materials to produce concrete. Lately, it has been shown that waste garnet can improve the durability of concrete if incorporated in appropriate proportions. The information generated in the literature on the endurance properties of such concretes is not enough. Therefore, the following sections will give a comprehensive overview of the durability of various waste garnet-based concretes.

Drying Shrinkage

Drying shrinkage is a measure of the deformation of concrete when water is withdrawn, eventually causing contraction. It normally leads to cracking and subsequently creates durability problems (Bissonnette et al., 1999). The property of drying shrinkage can be classified as deformation of concrete due to the fact that it causes changes to the dimensional configuration of the concrete specimen or member. It is considered to be a common phenomenon associated with cement-related products that induces reductions in volume due to the loss of moisture (Rao, 2001; Holt and Leivo, 2004; Al-Saleh and Al-Zaid, 2006). Feng et al. (2011) investigated the drying shrinkage of concrete prepared with iron mine tailings. Iron mill tailing sand, iron mill stone, natural sand, and crushed limestone were used as aggregates to produce concretes of grade C30 and C60. It was concluded that whether C30 or C60, the drying shrinkage of concrete made from tailing sand and tailing stone is lower than that of concrete made from natural sand and common crushed limestone. Similar findings were reported by Zhang (2013) that the drying shrinkage of mixed-sand concrete is lower than that composed of river sand. Furthermore, with increasing iron tailing proportions, the drying shrinkage of mixed-sand concrete was found to decrease. This was ascribed to the filling of the concrete matrix pores by the iron ore tailings, which contribute to the optimization of the porous structure and thereby enhance the density and strength of the hardened concrete.

Bai et al. (2005) examined the strength and drying shrinkage properties of concrete composed of FBA as fine aggregate. It was found that at a constant w/c of 0.45 and 0.55, the drying shrinkage values of all the prepared concretes were inferior to the control concrete at all curing ages, except for concrete prepared with a 50% replacement level at 0.55 w/c. Khatib and Ellis (2001) inspected the effects of the partial substitution of fine aggregates with waste foundry sand of three different types on the drying shrinkage properties of concrete. These types of foundry sands include fine white sand in the absence of clay and coal, foundry sand prior to casting (blended), and foundry sand after casting (waste). The level of partly substituted natural sand was varied from 25% to 100% to determine the drying shrinkage of concrete up to 60 days of curing. Experimental results revealed the following: (1) a lower concrete expansion for white sand concrete than the blended one at a low sand substitution level of 25%, (2) an increase in the volumetric change of concrete due to the substitution of standard sand with the aforementioned three types of sands, (3) higher drying shrinkage values for concrete composed of waste sand and lower values for those containing white sand. The volume change or shrinkage in concrete is a major concern for the construction industries, wherein strength and durability are greatly affected by shrinkage. Drying shrinkage, being the major deformation due to the withdrawal of water from concrete, often causes contraction, leads to cracking, and eventually makes the concrete non-durable (Khatib, 2001).

Concretes Resistance to Carbonation

Carbonation occurs due to the chemical reaction of the hydroxides present in the cement paste with the surrounding CO_2. The carbonation of concrete is one of the main reasons for the corrosion of steel reinforcements and is also one of the major factors in the degradation of concrete structures (Khunthongkeaw, 2006). During cement hydration, large amounts of calcium hydroxide (CH) are released, thereby increasing the pH above 12.5 in the concrete's aqueous phases. These conditions are usually associated with the development of passivated oxide layers surrounding the reinforcement steel, which protect it from corrosion (Valcuende and Parra, 2010). The atmospheric CO_2 enters the concrete's porous networks and undergoes a moisture-mediated reaction with $Ca(OH)_2$ and hydrated calcium silicates. Furthermore, CH and hydrated calcium silicates are converted to $CaCO_3$ (Equation 2.1). In this process, the hydroxide concentration in the pore solution decreases and the concrete pH drops below 9 (Bakharev, 2003). This action destroys the passivated oxide layers and promotes the corrosion of the steel reinforcement in the presence of water and oxygen.

$$Ca(OH)_2 + H_2O + CO_2 \rightarrow CaCO_3 + 2HO \tag{2.1}$$

According to Neville (2011), the carbonation rate is usually decided by the concrete's moisture content, relative humidity (RH) in a given environment, temperature, and the size of the specimen. One of the foremost factors that influences the concrete's carbonation process is the material's permeability related to its porosity (Gonen and Yazicioglu, 2007). Therefore, the higher the total porosity, the larger the carbonation depth. Conversely, the carbonation process is slower when the porous structures are finer. An evaluation of the durability of concretes containing other waste as aggregates displayed no more significant effects on carbonation than those which occur with natural aggregates.

Evangelista and De Brito (2010) showed that the carbonation rate of concrete was lowered due to the incorporation of fine recycled aggregate in the concrete. Helene (2004) reported that the carbonation depth of concrete was reduced due to the increase in recycled fine aggregate content as a substitution for natural aggregates. Moreover, for 100% recycled aggregate substitution, the carbonation depth was observed to be lower than the control mix. This reduction in the carbonation depth was attributed to the enhanced alkalinity of the mortar, where recycled fine aggregates played a significant role in protecting the concrete surface from further carbonation. Zhao et al. (2017) reported a high-strength, high-performance concrete exposed to 10 years of outdoor conditions and observed that concrete containing FA had a negligible carbonation effect. In addition, Shi (2009) studied the effects of a mineral admixture on the mechanical strength and carbonation depth of high-performance concrete. The findings of this study revealed that class-F FA content above 30% was not advantageous to the improvement of the concrete's carbonation depth.

Concrete Resistance against Acid Attack

Acid mainly attacks concrete structures such as foundations, retaining walls, pavements, embankments, and bridge decks by dissolving the constituent materials, leading to a weakening of the affected concrete. Generally, concrete structures in industrial areas are vulnerable to corrosion because of acid rain, where sulfuric acid is the main factor (Nematzadeh, 2017). The impacts of sulfuric acid on the concretes are more damaging than sulfate attacks. Besides attacks by sulfate ions, the effects of dissolution due to hydrogen ions is also detrimental (Tahri and Abdollahnejad, 2017). The main attacking agent of acid rain water is mainly comprised of sulfuric and nitric acid (Neville, 2011). The pH values of acid rain water occur in the range of 4.0 to 4.5, which is responsible for the surface weathering of the exposed concrete structures (Neville, 2011).

Duan (2015) investigated the long-lasting properties and microstructures of fluidized-bed FA and metakaolin-based GPCs by exposing them to increased temperatures and varying acid attacks. These GPCs were produced by blending FA and metakaolin activated via sodium silicate and NaOH solutions. The synthesized specimens were further cured under microwave environments with heat different curing period. It was found that the GPCs changed negligibly in appearance after 28 days of captivation in acid solution. Furthermore, the change in mass in such GPCs was only 0.7%. The resistance of concrete to acid attack can be greatly improved by incorporating hydrophobic additives, polymer emulsion, or pozzolanic additives

(Monteny, 2000; Bassuoni and Nehdi, 2007, Shafiq et al., 2017). An investigation by Bakharev et al. (2003) on alkali-activated slag concrete (AASC) shows that AASC performed better than OPC when exposed to acetic acid. However, there is limited information on concrete with garnet waste exposed to acid media.

Resistance of Concrete to Sulfate Attacks

Generally, sulfate attacks occur when the cement reacts with sulfate-containing solutions, including natural or polluted ground water. In OPC, such attacks lead to a loss of mechanical strength, expansion, surface layer spalling, and finally breakdown (Žarnić et al., 2001). Meanwhile, inorganic alkaline polymers or GPCs such as alkali-activated metakaolin and FA show excellent tolerance to normal sulfate and sea water attacks due to having lower calcium phases. Bakharev (2005a) mentioned that a continuous cross-linked polymer structure of alumino-silicate is formed when NaOH is used as an alkali activator. Rangan (2008) pointed out that no damage occurs to the surface of FA-based GPCs when sodium silicate and NaOH are used as the alkaline activators. The specimen was immersed in sodium sulfate solution for about 1 year before conducting the test.

The discrepancies in NaOH or sodium silicate are ascribed to the different Si/Al ratios produced and to the larger or smaller quantities of (zeolitic) crystalline phases present in the matrix. In addition, the existence of silicate ions leads to the creation of more dense structures, with gels richer in Si (Duxson et al., 2005; Fernández-Jiménez and Palomo, 2005b; Fernández-Jiménez, 2006; Criado, 2007). Sulfate attacks in OPC-based concretes or mortar are mostly ascribed to the generation of expansive ettringite (AFt phase) and gypsum. The sulfate ions interfere inside the concrete and react with portlandite Ca $(OH)_2$ to form gypsum. In the existence of enough sulfate, the metastable mono-sulfo-aluminate converts into ettringite, which subsequently absorbs moisture to undergo expansion cracking and spalling (Žarnić et al., 2001). Generally, GP products are devoid of $Ca(OH)_2$ where mono-sulfo-aluminates are formed from resource materials that enclose little or no Ca. Finally, upon exposure to sodium sulfate solution, the growth of gypsum and ettringite ceases. This in turn results in the expansion of the matrix, implying that the GP samples have become resistant against sulfate attack.

Performance of Concrete at Elevated Temperatures

As interest grows in the application of spent garnet concretes for sustainability and environmental protection, it is also necessary to examine other properties such as fire resistance. Hence, the risk of exposure to elevated temperatures during fires cannot be ruled out. Concrete structures might be subjected to high temperatures and pressures throughout their performance for significant time durations. These structures include control containers, chimneys, pressure vessels, nuclear reactors, hot water containers, crude oil storage tanks, coal gasification structures, vessels for liquefaction in petrochemical industries, structures used in the coal and coke industries, furnace walls, foundations for blast furnaces in industrial chimneys, and aircraft runways (Arioz, 2007; Demirel and Keleştemur, 2010; Ismail, 2011; Neville, 2011).

In addition, other concrete structures such as walls and pipes may be significantly damaged due to exposure to elevated temperatures. Consequently, in order to determine the properties of such structures after exposure to high temperatures, it is necessary to evaluate the key hardening properties. The types of materials used in concrete have a significant impact on the characteristics of the concrete produced. It is decided by the aggregate and cement composition, porosities, moisture content, thermal composition and the sizes of the concrete components (Akçaözoğlu, 2013). Georgali and Tsakiridis (2005) reported that when concrete is exposed to elevated temperatures, it usually changes color with the cracking and spalling of the surfaces. This damage is evaluated based on visual expansion. Characteristics including elastic modulus, mechanical strength, surface texture, density, and volume distortion decreased drastically upon exposure to elevated temperatures. Sudarshan (2017) reported the impact of fire on the mechanical properties of concrete containing marble waste. Test specimens were exposed to fire at temperatures of 200°C, 400°C, 600°C, and 800°C in a gas-fired furnace. The results show that the higher temperature range (up to 800°C) also deteriorates concrete and reduces strength. The degradation of concrete at elevated temperatures is mainly due to mechanical and chemical changes in the aggregate and cement paste. Al Qadi (2014) investigated the effects of fiber content and specimen shape on the residual strength of polypropylene fiber SCC exposed to elevated temperatures at 200°C, 400°C, and 600°C. The presence of fibers at different dosages in concrete does not affect the compressive strength at 200°C and 400°C, while it considerably increases the residual compressive strength of concretes after exposure to 600°C. Fares (2010) reported the performance of SCC subjected to high temperatures. Two mixtures of SCC and one of vibrated concrete were used. Specimens were heated at different temperatures (150°C, 300°C, 450°C, and 600°C) with 1 hour of exposure. The results shown that spalling happens at 315°C. Moreover, compressive strength decreases with an increase in temperature.

These consequently lead to a reduction in the structural stability of the concrete structures (Arioz, 2007). In addition, at high temperatures the chemical and physical properties of the concrete structure are considerably affected (Demirel and Keleştemur, 2010). This is because at temperatures higher than 110°C, the cement hydrates decompose when $Ca(OH)_2$ is ruptured and $CaCO_3$ undergoes decarbonation. These aggregates also undergo alterations that lead to the usual loss of structural integrity (Demirel and Keleştemur, 2010), and at temperatures above 300°C, microcracks are generated through the materials (Kodur et al., 2003). Previous works have examined the influences of heating rate, maximum exposure temperature, and time on the residual strength of concretes prepared using varying mix proportions (Demirel and Keleştemur, 2010). It has also been reported that the method used to overcome the spread of fire (e.g., water, air, and chemical cooling) during an outbreak plays a vital role on the residual strength of the concrete (Lin, 2011; Cree, 2013).

LEACHING BEHAVIOR OF CONCRETE

The concentration of heavy metals in processed waste is a source of concern. Several researchers have shown that hazardous metals in waste have been used to solidify and effectively stabilize cement paste. Park (2000) examined the mobility of toxic materials in OPC, quick-setting agent (QSA), and clinker kiln dust (CKD). It was indicated that a lower leaching of heavy metals was observed in the mix. Lin et al. (2003) reported that at 0.38 w/c ratio, the mobilization of heavy metals from 10% to 40% cement substitution by FA and slag in a mortar sample was below the regulatory standards for hazardous substances set by the U.S. Environmental Protection Agency. Another study by Giergiczny and Krol (2008) revealed the effective mobilization of heavy metals with the use of OPC, FA, and GGBFS in the blended mortar mixtures. Shi and Kan (2009) reported that heavy metals leaching cement paste containing FA were below the Chinese regulatory limits. Choi et al. (2009) examined the leaching of toxic materials in mixtures of binary blends of tailing waste from tungsten and GGBFS in mortars.

It was shown that mine tailings contain high concentrations of heavy metals. However, to fill the information gap on the leaching behavior of garnet waste material, further research is needed.

SUMMARY OF RESEARCH

1. This chapter has depicted in detail the relevant literature on the effective utilization of waste materials in the construction sectors.
2. The existing literature has been critically evaluated to determine the feasibility of using waste materials such as manufacturing sand, bottom ash, slag, iron ore tailings, and recycled aggregates as sand replacement to create good-quality concrete.
3. No case studies on utilizing this spent garnet have been found in the literature. However, the preparation and characterization of SCGPC made from waste materials in place of river sand are prerequisites to reaching the target.
4. Some literature hinted that the effective utilization of industrial and domestic wastes would be not only economically viable but also a step toward environmental protection.
5. It has been emphasized that the problems of environmental degradation and the depletion of natural sand necessitate the use of waste material as fine aggregate in the production of concrete. Several works found that concretes made from recycled waste materials such as FA, slag, manufacturing sand, bottom ash, and recycled fine aggregate have desirable properties suitable for construction purposes. Despite of great abundance of spent garnet worldwide, it has not been researched thoroughly to identify its potential for making diverse concrete structures. The fresh, hardened, durability, and microstructural characteristics of such concretes need careful scrutiny. In this respect, the present work is highly significant and will shed some light on how to meticulously reuse huge amounts of spent garnet to manufacture concrete structures.

3 Characterization of Raw Materials and the Fresh and Hardened Properties of Self-Compacting Geopolymer Concrete

CHARACTERIZATION OF MATERIALS

For this study, spent garnet was utilized as natural fine aggregate in place of river sand to prepare self-compacting geopolymer concrete (SCGPC) with improved attributes. The spent garnet was characterized based on its physical properties, chemical composition, and leaching behavior. It was also characterized based on its morphology and microstructure. The results for these properties are depicted in the following sections.

PHYSICAL ANALYSIS OF SPENT GARNET

The spent garnet showed a specific gravity value of 3.0, which is higher than that of natural sand (which usually varies from 2.6 to 2.7) (Neville, 2011). The high specific gravity of spent garnet was ascribed to the elevated content of iron oxide (Fe_2O_3). The water absorption value of 6% was higher than the maximum limit of 3% recommended by BS EN 1097-6 (2013) for fine aggregate. The bulk density was 1922 kg/m^3, which is higher than the limits of 1300 to 1750 kg/m^3 required for normal-weight concrete as specified by Mehta and Monteiro (2013) and Neville and Brooks (2010). The bulk density of spent garnet is higher when compared with that of river sand, and this might be due to reduced void content. The finer particles of spent garnet might have filled pores and optimized the pore structure; however, it leads to a higher water demand (Li, 2013). Figure 3.1a,b shows the results of field emission scanning electron microscopy (FESEM) for spent garnet and natural river sand, respectively. FESEM analysis showed that both of them had an angular shape. The spent garnet had a porous and highly rough surface with an irregular shape that was well dispersed, while the natural river sand had a smoother surface. The nature of the particle packing and the surface texture of spent garnet could contribute positively to the water demand and therefore cause low concrete workability. Similar behavior has been reported by Zhao (2014).

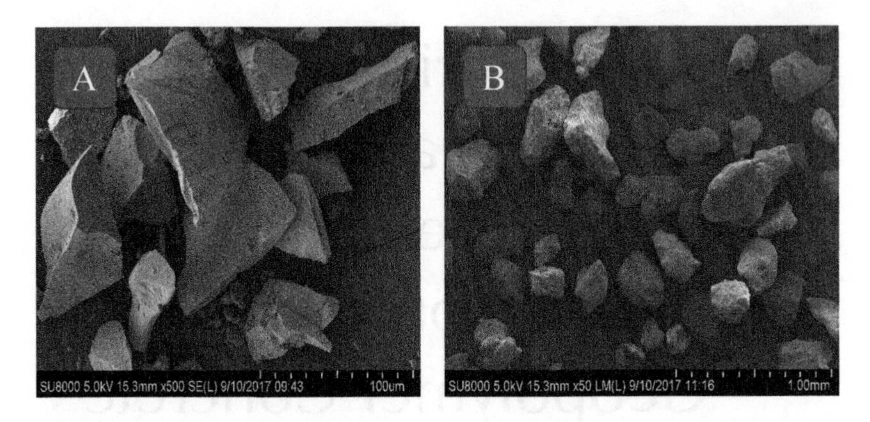

FIGURE 3.1 FESEM for (a) spent garnet, (b) natural river sand.

Grading of Spent Garnet

Figure 3.2 shows the results of the grading of spent garnet and river sand as derived from sieve analysis. The coefficient of grading was 1.02 for both spent garnet and river sand, which were greater than 1. Therefore, the spent garnet material was also considered well graded. The particle size distribution of the spent garnet material falls within the upper and lower limits of medium grading (M) for fine aggregate specified by BS 882 (1992), which is equivalent to the zone II classification of BS 812-103.1 (2011). For river sand, this fell into the coarse grading limits (C) of BS 882 (2011), which falls into the zone I classification of BS 812-103.1 (2011). Fine aggregates for concrete are generally required to conform to any one of the three grading limits of BS 882 (2011). The limits in which the grading falls were designated as coarse (C),

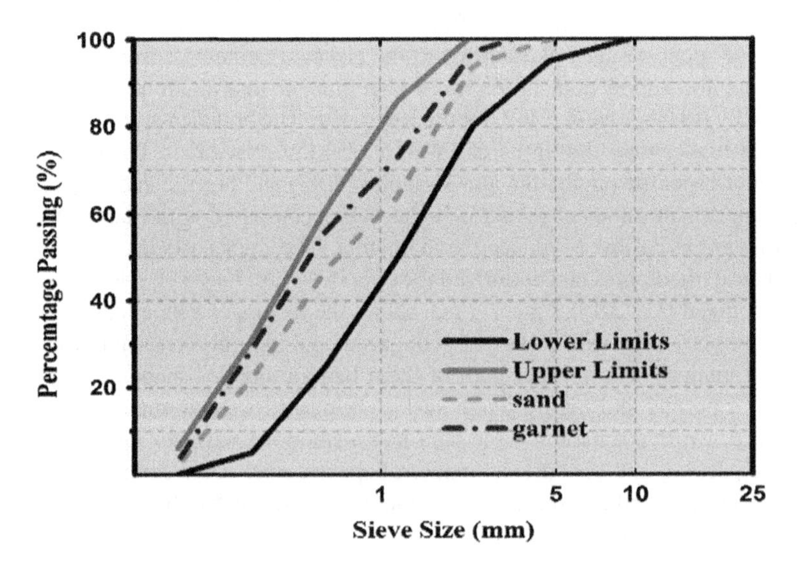

FIGURE 3.2 Sieve analysis results of spent garnet and river sand.

medium (M), and fine (F), respectively. The percentage retained at 5 mm sieve size was 0% for spent garnet and 2% for sand (raw data in appendix), which were within the recommended limits of 0%–5% of the BS 882 (2011) grading requirements for fine aggregates. This confirmed that the grading of spent garnet conformed to the recommended grading for fine aggregates suitable for concrete production.

Chemical Composition of Spent Garnet

The chemical compositions of the spent garnet and the river sand were evaluated using X-ray fluorescence (XRF) spectroscopy, as listed in Table 3.1. The major component of the spent garnet was iron oxide (Fe_2O_3). The mass contents of iron oxide of the spent garnet and river sand were 43.06% and 0.7%, respectively. The alumina (Al_2O_3) content of the spent garnet was 13.88% and the silicon oxide (SiO_2) content was 33.76%, while the sand had SiO_2 of 96.4%. Iron oxide (Fe_2O_3) was the main colorant in the spent garnet, being responsible for the reddish color. The calcium oxide (CaO) content for the spent garnet was 4.15% and 0.14% for the river sand. However, the presences of CaO, ZnO, and P_2O_5 in the chemical composition will possibly affect the hydration process of SCGPC (Giergiczny and Krol, 2008). Interestingly, no trace of sulfur trioxide was found in the chemical composition. The reaction between alkaline solution, silicon (Si), and aluminum (Al) in the spent garnet probably affected the geopolymer paste. The spent garnet's chemical composition showed high ratios of silicon (Si) and aluminum (Al) up to 47.64%.

Ground Granulated Blast Furnace Slag

The presence of enriched alumino-silicate and the economic and abundant nature of slag makes it suitable for the manufacturing of SCGPC. Table 3.2 summarizes the chemical compounds present in the slag. Ground granulated blast furnace slag (GGBFS) has strong alkali solutions, sodium hydroxide (NaOH), and soluble

TABLE 3.1
Chemical Composition of Spent Garnet

Chemical Compounds	Weight % of Spent Garnet	Weight % of Sand
Fe_2O_3	43.06	0.7
SiO_2	33.76	96.4
Al_2O_3	13.88	–
CaO	4.15	0.14
MgO	2.91	–
MnO	1.08	–
TiO_2	0.78	1.1
K_2O	0.14	–
P_2O_5	0.21	–
ZnO	0.06	–
Cr_2O_3	0.05	–

TABLE 3.2
Chemical Compounds Present in GGBFS

Chemical Compounds	Composition (%)
SiO_2	33.80
Al_2O_3	13.68
Fe_2O_3	0.4
CaO	43.2
MgO	0.46
K_2O	0.21

silicates, where the dissolved Al_2O_3 and SiO_2 species undergo geopolymerization to form a strong three-dimensional amorphous alumino-silicate network.

Material Safety

GMA Garnet Pty Ltd. prepared the material safety data sheets (MSDS) following the National Occupational Health and Safety Commission (NOHSC) 2003 standards.

Hazards Identification

Aspects of the materials were identified according to the NOHSC: 1008 (2004) standard, whereas supplied products were found to contain traces of quartz (crystalline silica). Table 3.3 shows the hazards identification. Upon abrasive usage, these products can split into a breathable size of dust particle. It is documented that respirable/breathable crystalline silica is carcinogenic and causes silicosis and cancer. In the present work, the product being predominantly garnet meant it was a harmless substance. The presence of fine powder in the abandoned product was mostly in the form of calcium carbonate, which is also safe and non-hazardous. The following identification indices were used: (1) Risk Phrase (T R49): containing silica crystals that may create cancer due to inhaling; (2) Safety Phrase (S22): no breathing of dust that is released from the used product.

TABLE 3.3
Hazards Identification

A: Firefighting Measures	
Flammability	Non-flammable
Flashpoint	Non-explosive
General hazard	Non-flammable
B: Accidental Release Measures	
No special precautions necessary.	
C: Handling and Storage	
No special precautions necessary.	

First Aid Measures

It is proven that short or enduring exposure to spent garnet does not cause any severe health problems to workers. However, the breathing of crystalline silica dust even at very low concentration may cause silicosis and cancer. The following factors were considered prior to the use of spent garnet.

- Swallowed: Non-toxic without any health hazards for accidental ingestion in small amounts. However, special medical care may be required for large amounts of ingestion, which may result in irritation.
- Eye: Particle and dust exposure that causes eye irritation requires thorough cleaning by water to remove the dirt completely; persistent irritation will require medical assistance.
- Skin: Skin contact with these particles at normal pressure does not cause any health hazards. However, contact at high pressure may cause skin damage by abrasion. Medical care is required in case any wounding or skin damage occurs.
- Inhaled: Contact with dust created by blast-cleaning medium causes throat and lung irritation, coughing, or shortening of breath.

Leaching Analysis

The contents of heavy metals such as arsenic, cadmium, chromium, cuprum, nickel, barium, lead, selenium, and cobalt in the raw garnet were tested to guarantee the material's safety. The Toxicity Characteristic Leaching Procedure (TCLP) established by the U.S. Environmental Protection Agency (EPA) was employed to determine the heavy metal contents of the garnet, as summarized in Table 3.4. The heavy metal çontents were discerned to be lower than the control limits. The contents of As, Ba, Cr, Cd, Znm, and Pb in the leachate were below the allowable regulatory

TABLE 3.4
Presence of Heavy Metals in Spent Garnet Determined Using TCLP Analysis

	Results (mg/L)	
Detected Metals	**Spent Garnet**	**Regulatory Limits**
Arsenic (As)	0.000103	5.0
Cadmium (Cd)	0.000445	1.0
Chromium (Cr)	0.006195	5.0
Cuprum (Cu)	0.00462	25
Nickel (Ni)	0.000159	2
Barium (Ba)	1.145093	100
Lead (Pb)	Not Detected	0.05
Selenium (Se)	0.000757	1.0
Cobalt (Co)	0.00017	8
Zinc (Zn)	0.127609	250

limit for waters (ground water, surface water, and tap water) and farmland soil based on the U.S. EPA (2000) standards. Thus, the spent garnet used in this study can be considered safe material with the following specifications.

MICROSTRUCTURE CHARACTERIZATION OF SPENT GARNET AND RIVER SAND

The morphology and microstructure of the river sand and spent garnet were investigated through X-ray diffusion (XRD), thermogravimetric analysis (TGA), and Fourier-transform infrared (FTIR) spectroscopy. The results are hereby presented and discussed.

PHASE ANALYSIS

Figure 3.3 displays the XRD patterns of spent garnet, which revealed several sharp crystalline peaks in the angular range of 25° to 100°. The observed peaks were assigned to the crystalline phases of quartz (SiO_2), magnesium (Mg), and aluminum cobalt. Quartz is the major phase and is generally known to be non-reactive but usually increases the packing density and long-term strength development. The significance of the results is the ability of geopolymer paste to respond properly with sand and spent garnet; consequently, calcium aluminum silicate hydrate (C-A-S-H) gel develops in increments.

Figure 3.4 depicts the XRD patterns for river sand, where the appearance of crystalline peaks indicates the presence of quartz (SiO_2) as the major crystalline mineral. The crystalline silicate phases could be beneficial for generating calcium silicate hydrate (C-S-H) gel or C-A-S-H in the concrete. In the C-A-S-H geopolymerization process, solid alumina-silicate is transformed into synthetic alkaline alumina-silicate. An aluminate and silicate species are produced due to the dissolution of the solid alumina-silicate via hydrolysis of alkali solution. The species are activated because of the high-solution pH. First, the super-saturation of alumina-silicate gives

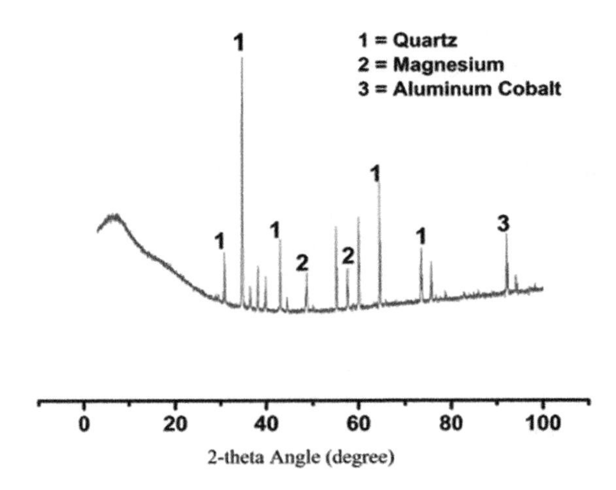

FIGURE 3.3 XRD pattern of spent garnet.

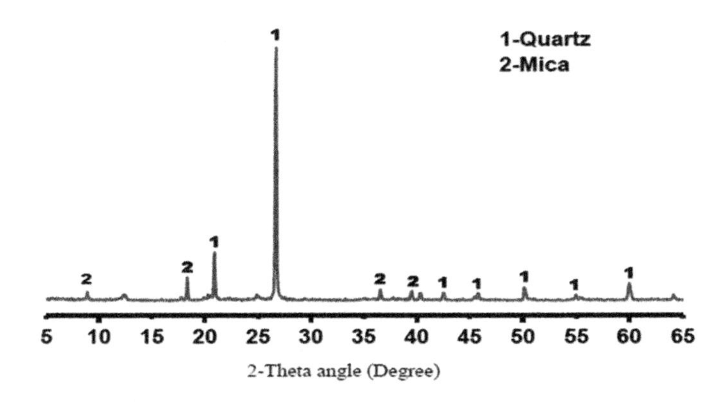

FIGURE 3.4 XRD pattern of river sand.

rise to a gel, which further forms a large network of oligomers via condensation. In the formation of geopolymers, water is released that is further excluded from the geopolymer matrix during the curing and drying process. Consequently, discontinuous nanopores are left over within the geopolymer matrix, which in turn enhance the performance of geopolymer.

Thermal Analysis

TGA is a thermal analytical process in which the mass of a particular sample in a controlled environment is continuously measured as a function of temperature for a defined temperature program. Figures 3.5 and 3.6 illustrate the TGA/differential thermal analysis (DTA) curves of spent garnet and river sand, respectively.

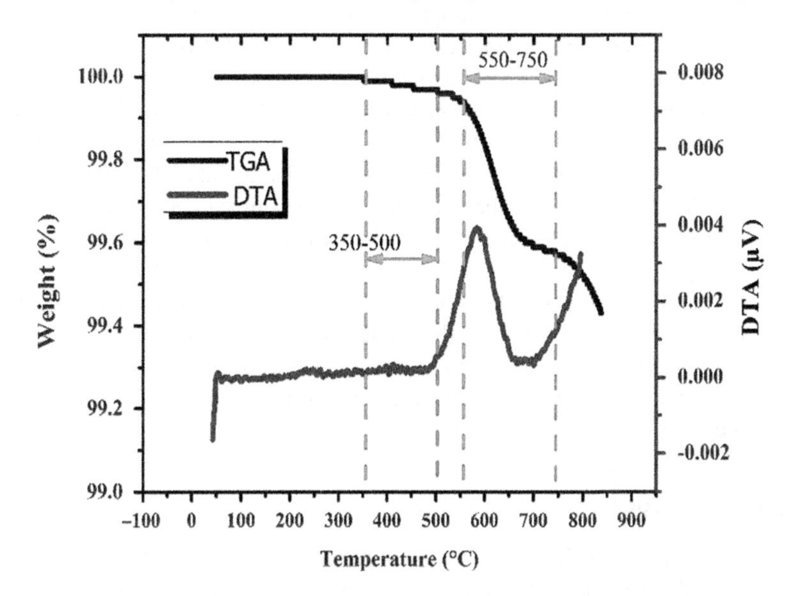

FIGURE 3.5 TGA/DTA curve of spent garnet.

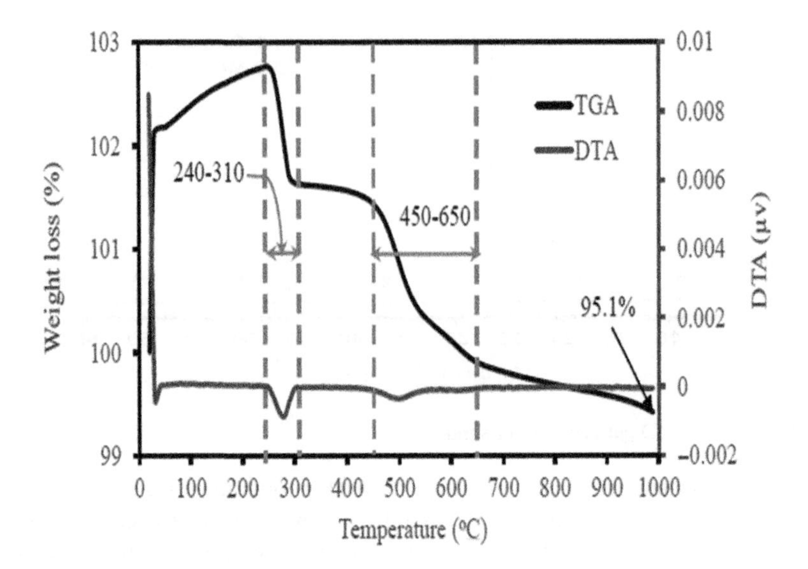

FIGURE 3.6 TGA/DTA curve of river sand.

The endothermic dip in the spent garnet was observed between 350°C and 500°C. This dip is usually related to the loss of surface water and the dihydroxylation of the material. The exothermic peak (heat release) appeared between 550°C and 750°C due to the oxidation reaction of the organic matter in the spent garnet, with a weight loss of 0.4% in this period. Meanwhile, the oxidation of hematite also occurred in this temperature range, but the heat and weight variations were concealed by the stronger oxidation reaction of the organic matter in the spent garnet. The observed weight loss of about 0.2% for the sample in between 700°C and 900°C was ascribed to the decomposition of calcite into CaO at 768.5°C (Cai, 2016) and the dehydroxylation of quartz in the range of 850°C to 900°C.

The endothermic peaks of river sand at 40°C and 130°C were also attributed to the loss of surface water and dihydroxylation. The difference between the DTA temperatures due to the endothermic reaction of the sand and that of the spent garnet was significant. The TGA curve further showed that the sample decomposed from 100% to 99.4% for spent garnet, which corresponds to the mass loss of 0.4% and 0.6% due to the decomposition of adsorbed water at 450°C–550°C and 700°C–850°C, respectively, making a loss of about 1%. However, river sand decomposed from 100% to 95.1%, a loss of 4.9%. Comparing the values, spent garnet decomposed less than river sand, which could be due the loose nature of the particles. However, both sand and spent garnet showed good properties in high temperatures, which can be important for the fire resistance of concrete.

Bonding Analysis

FTIR spectra provide valuable information regarding the molecular bonding vibration via the absorption or transmission of solids and liquids, mainly. Figures 3.7 and 3.8 present the FTIR spectra for spent garnet and river sand, respectively. The

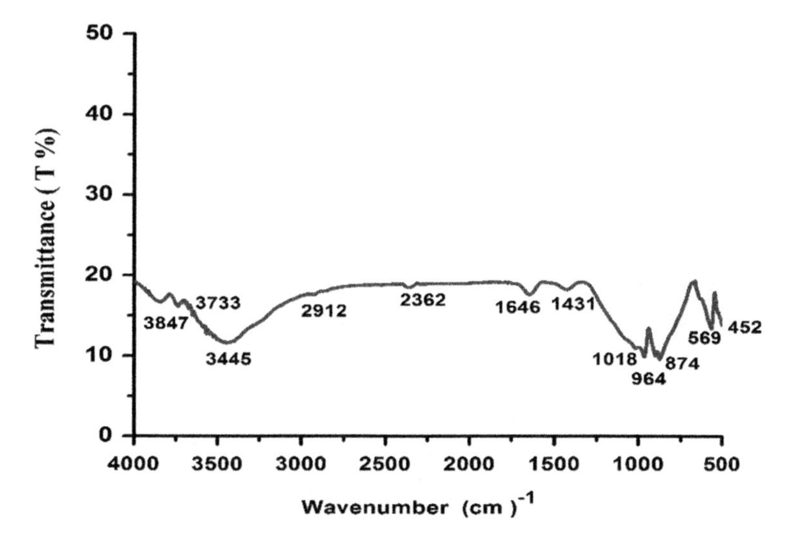

FIGURE 3.7 FTIR spectra of spent garnet.

spectra of spent garnet (Figure 3.7) were comprised of two groups of absorptions such as the outer hydroxyl groups and the inner hydroxyl group. The external groups were located in the upper unshared plane, whereas the inner groups were situated in the lower shared plane of the octahedral sheet. The spectrum of the minerals was arranged in the sheets as per the occupancy of octahedral and tetrahedral sites in the lattice. The inner hydroxyl groups were positioned amid the tetrahedral and octahedral sheets, which revealed an IR absorption of ≈3733 cm^{-1}. The strong absorption band observed at about 3847 cm^{-1} was allocated to the in-phase symmetric

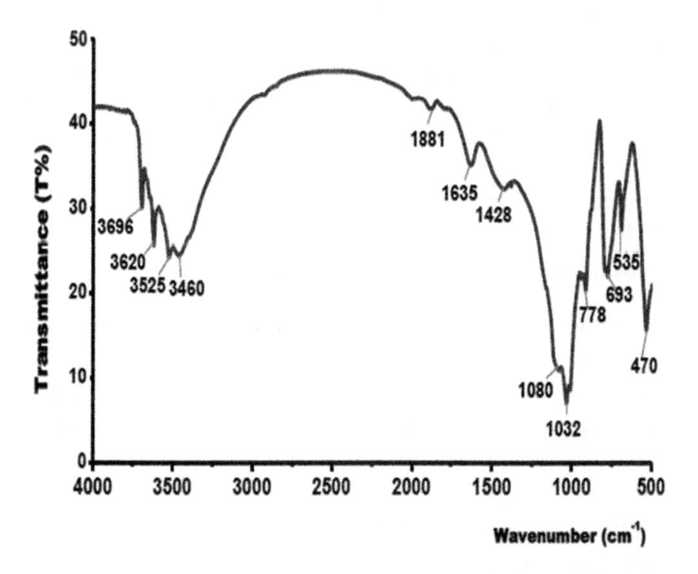

FIGURE 3.8 FTIR spectra of river sand.

stretching. Conversely, the occurrence of a weak band at 3445 cm^{-1} was allocated to the out-of-plane stretching vibration (Kristof, 1997). The absorption bands at 1018 and 964 cm^{-1} were allocated to the Si-O symmetric and asymmetric stretching vibrations, respectively (Kaufhold, 2012). The IR absorption bands at 569 and 452 cm^{-1} were allocated to the Si-O-Al and Si-O-Si bending vibrations, respectively.

The band at 1431 cm^{-1} exhibited the stretching vibration of Fe_2OH, a characteristic compound of spent garnet minerals, as also revealed in the XRF spectra. The IR spectra spent garnet further showed that the absorption bands were due to the bonding vibrations of the OH and Si-O groups. The presence of these groups often plays a significant role in the water absorption rate and the differentiation of the mineral composition. The FTIR spectra of river sand (Figure 3.7) revealed significant IR peaks at 778 and 1032 cm^{-1}, which were allocated to the symmetric and asymmetric stretching vibrations of the Si-O bond, respectively. Yet again, the appearance of an IR band at 694 cm^{-1} was allocated to the asymmetric and symmetric bending vibrations of the Si-O bond. The bands at 3620 and 3697 cm^{-1} were allocated to the symmetric and asymmetric stretching of the calcium hydrate (CH) group (Bera et al., 2013). The bands associated with the stretching vibrations of the OH group of interlayer water molecules during surface absorption were also in evidence. It was affirmed that pure silica was the main component of the river sand. The FTIR results were supported by the XRF data, both of which revealed the high silica content of the river sand (Table 3.5).

FRESH PROPERTIES OF SCGPC

EFFECTS OF SPENT GARNET ON THE FRESH PROPERTIES OF SCGPC

Table 3.6 summarizes the fresh properties of the prepared SCGPC specimens obtained by slump flow, L-box, and V-funnel analyses. The mix sample M1 was selected as the control sample, whereas the M2, M3, M4, and M5 specimens were chosen as the spent garnet samples. Quantitative and qualitative analyses indicated that the fresh properties of the concrete mixes conformed to the EFNARC limits of SCGPC (EFNARC, 2002). All the aforementioned test results of the proposed

TABLE 3.5
Functional Groups of Bonding Analysis

Wave Numbers (cm^{-1})	Bond	Standard
3847–3733	In-phase symmetric stretching	*PR
3445	Out-of-plane stretching vibration	*PR
1018–964	Si-O	*PR
569	Si-O-Al	*PR
1431	Si-O	*PR
778–1032	Si-O	*PR

*PR: Previous research of standard method

TABLE 3.6
Fresh Properties of Prepared SCGPC

Mixes	Slump Flow (mm)	T50 cm Slump Flow (sec)	V-Funnel (sec)	L-Box Ratio (H_2/H_1)	SP (%)	Molarity (M)	Remark as per EFNARC Standard
M1	671	5.5	12	0.91	3	8	OK
M2	675	5	11.5	0.92	3	8	OK
M3	681	4.5	11	0.93	3	8	OK
M4	692	4	7.5	0.95	3	8	OK
M5	700	3.5	6.5	0.97	3	8	OK
Minimum	650	2	6	0.8	Acceptance criteria as per EFNAR		
Maximum	800	5	12	1			

SCGPC specimens revealed that the substitution of spent garnet for river sand could significantly enhance their workability. Table 3.6 reveals the slump flow values of 671, 675, 681, 692, and 700 mm respectively. Similarly, it can be seen that the slump flow increased with the increase of spent garnet. This could be attributed to the irregular shape and higher water absorption of the 6% spent garnet material, as discussed in the characterization. However, when the percentage of spent garnet increased, the T50 and V-funnel values decreased, the T50 and V-funnel tests being inversely proportional with the slump flow test values. The passing and filling capacities of SCGPC measured by L-box test show that an increased spent garnet percentage in the concrete decreased the L-box values. This could be attributed to the high ratio of fine particles in the spent garnet compared with the fine aggregate.

In comparison with the control, the slump flow of the SCGPC was found to increase by as much as 29% when the garnet level in the concrete mix was increased to 100%. The fresh SCGPC was less cohesive and less workable than the SCGPC consisting of spent garnet. This observation was ascribed to the overall morphology (shape and size) of the spent garnet, where the fresh state properties of the concrete were improved (Mehta, 1986). Moreover, the fineness modulus of garnet (2.05) was much lower than the minimum value (2.3) of sand as specified by the ASTM C33 standard. Thus, an increase in the extra water demand in the SCGPC caused an enhancement in the slump flow properties. The fresh slag-based SCGPC was easily controlled for up to 40 minutes, exclusive of any setting or compressive strength degradation. The use of super-plasticizers has made it possible to produce a workable SCGPC. The results indicate the optimum quantity of super-plasticizer to be 3%, as shown in Figure 3.9. The L-box values were also effected by the spent garnet percentage in the SCGPC samples, where an increased spent garnet percentage reduced the L-box values.

Density of Fresh SCGPC

The density of the SCGPC was largely decided by the unit weight of the mixture aggregates. Table 3.7 lists the fresh densities of the synthesized SCGPC. The fresh

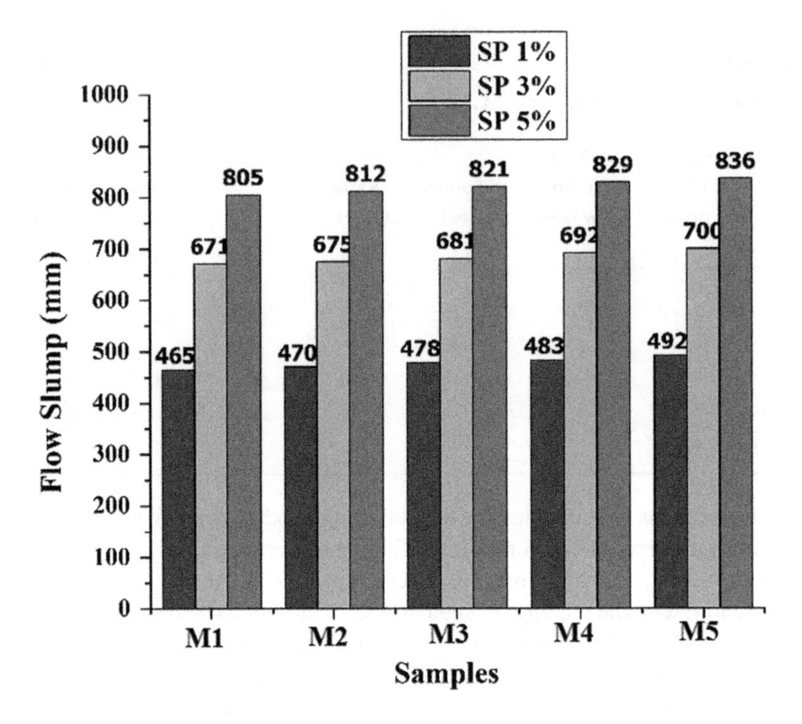

FIGURE 3.9 Effect of super-plasticizer on SCGPC flow slump.

TABLE 3.7
Fresh Density of Prepared SCGPC

Concrete Mixes	Fresh Density (kg/m^3)
M1	2380
M2	2405
M3	2420
M4	2464
M5	2493
BS EN 12350-6 limits	**2300–2400**

density increased as the spent garnet content in the concrete increased. It can be seen that, although the fresh density was lower in control specimen M1 compared with that of the spent garnet concrete (M2, M3, M4, and M5), in each case it is above the minimum of 2300 kg/m^3 specified by BS EN 12350-6 (2009). The higher density values of spent garnet concrete can be ascribed to the high concentration of iron in the material, which was heavier than river sand in terms of weight. Another factor could be that the spent garnet was finer than the natural sand, and the particles were angular and irregular, which has a better physical filling effect than the natural sand; this led to a higher packing density in the fresh SCGPC.

HARDENED PROPERTIES OF SCGPC

Density of Hardened SCGPC

Table 3.8 summarizes the measured dry densities of synthesized hardened SCGPC specimens at 28 days of curing. The density of the hardened concrete was measured as the ratio of average weight per unit volume using a simple calculation method specified in the BS EN 12390-7 (2009) standard. There was an increase in dry density with an increase in spent garnet content compared with control M1. The higher density values of the spent garnet concrete can be ascribed to the high concentration of iron in the material, which was heavier than river sand in terms of weight. Another factor could be that the spent garnet was finer than the natural sand. However, the density of control M1 was found to be 2370 kg/m^3, which ranged between 2300 and 2400 kg/m^3 for regular-weight concretes (BS EN 12390-7, 2009).

Compressive Strength of SCGPC

Table 3.9 summarizes the measured compressive strength for the control and the spent garnet-based concretes cured over different periods. The compressive strength of the garnet-based concrete was lower compared with the control for all curing durations. The reduction in the ratio of compressive strength (after 28 days) of the C2,

TABLE 3.8
Hardened State Density of SCGPC (28 Days)

Concrete Mixes	Density (kg/m^3)
M1	2370
M2	2395
M3	2410
M4	2445
M5	2471

TABLE 3.9
Curing Duration–Dependent Compressive Strength of Synthesized SCGPC Specimens

		Compressive Strength (MPa)		
Samples		3 days	7 days	28 days
0% garnet	C1	70.2	73.2	79.8
25% garnet	C2	71.6	73.1	78.2
50% garnet	C3	67.5	70.4	76.3
75% garnet	C4	65.9	69.1	75.4
100% garnet	C5	61.4	66.2	70.3

C3, C4, and C5 specimens compared with control C1 was discerned to be −2.04%, −4.41%, −5.50%, and −11.92%, respectively. The reduction observed in the compressive strength was largely attributed to the fineness of the spent garnet particles, which lacked the appropriate gradient and shape for filling and optimizing the pores. However, the coarse and angular texture of the spent garnet materials enhanced the bonding between the slag-aggregate interface, increasing the strength.

The compressive strength of the SCGPC was reduced with the increased garnet percentage, where the replacement of normal sand with spent garnet up to 100% was favorable for producing SCGPC due to the absence of strength degradation. However, SCGPC specimens cast with 25% spent garnet substitution for normal sand yielded the maximum strength at 28 days. Figure 3.10 displays the curing period-dependent development of the compressive strength of the SCGPC for C1 (100% sand). The increase in the compressive strength of the SCGPC was found to be proportional to the NaOH molarity and curing period at 28°C. The increasing NaOH molarity influenced the compressive strength of the SCGPC, which can further accelerate the geopolymerization process over a certain curing duration. The compressive strength at 12 M NaOH was higher by 15.8% than at 8 M. However, at a high NaOH molarity of 12 M, the workability of the SCGPC was reduced and the viscosity enhanced. An NaOH molarity of 8 M showed the best performance for SCGPC.

SPLITTING TENSILE STRENGTH

Figure 3.11 depicts the splitting tensile strength of the prepared SCGPC, which decreased with the increasing percentage of spent garnet. The splitting tensile

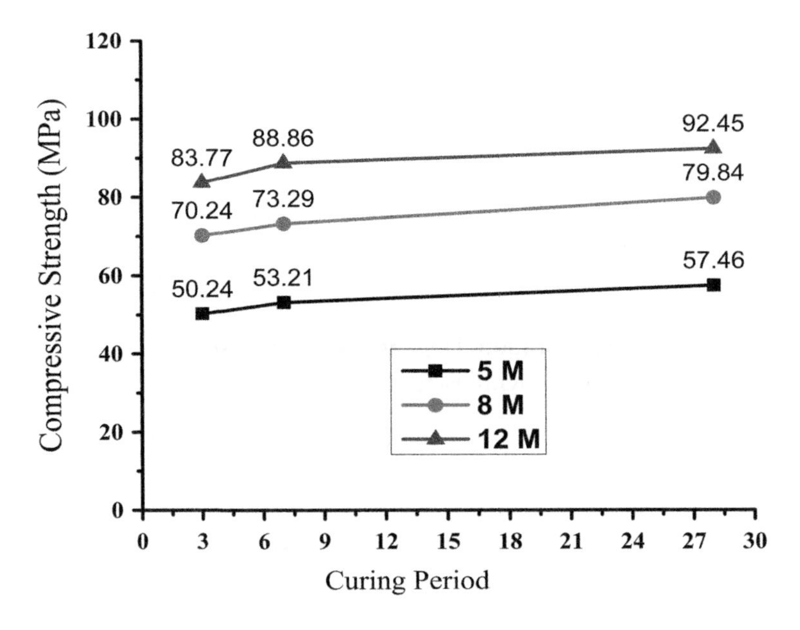

FIGURE 3.10 Influence of curing period and NaOH molarity on the compressive strength of SCGPC.

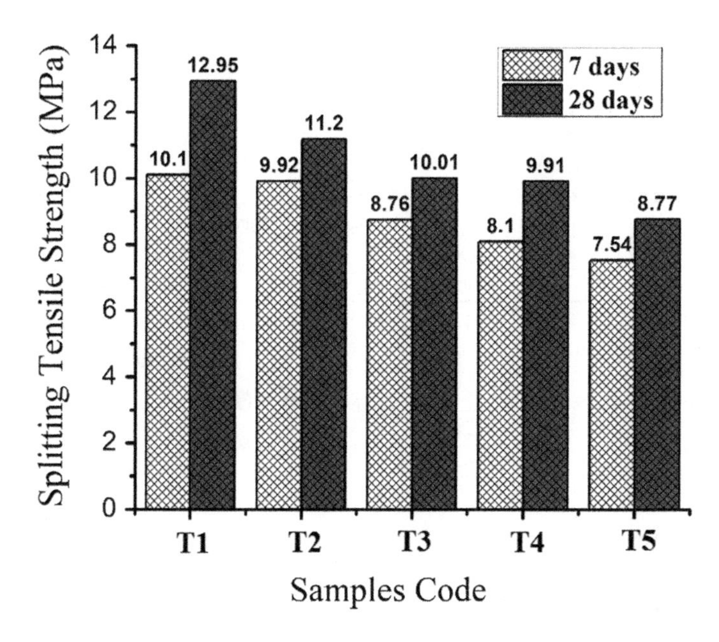

FIGURE 3.11 Curing time-dependent splitting tensile strength of various SCGPC specimens.

strength were reduced after 7 and 28 days of curing. Besides, a sudden declination in the splitting tensile strength of the SCGPC was observed. The values of the splitting tensile strength for the T2, T3, T4, and T5 specimens after 28 days dropped by −13.51%, −22.78%, −23.47%, and −32.28%, respectively, compared with T1 (control mix). The decrease in tensile strength with the increasing inclusion of spent garnet into the mix was due to the weakening of the bonds between the tiny spent garnet particles and the binder paste. Table 3.10 shows raw data for the splitting tensile strengths of various SCGPCs.

The splitting tensile strength of the SCGPC had a strong correlation with the compressive strength. The relationship between splitting tensile strength and compressive strength is shown in Figure 3.12. The correlation was determined for compressive strength in the range of 70.32 to 79.84 MPa and splitting tensile strength in

TABLE 3.10
Raw Data for Splitting Tensile Strength of Various SCGPCs

Sample		7 days			28 days		
0% garnet	T1	10.30	9.70	10.40	12.55	12.41	13.90
25% garnet	T2	10.35	9.81	9.60	10.95	11.11	11.42
50% garnet	T3	9.52	8.84	7.93	9.65	10.02	10.36
75% garnet	T4	7.79	7.91	8.65	9.51	10.22	10.01
100% garnet	T5	7.40	7.25	7.97	8.36	9.01	8.94

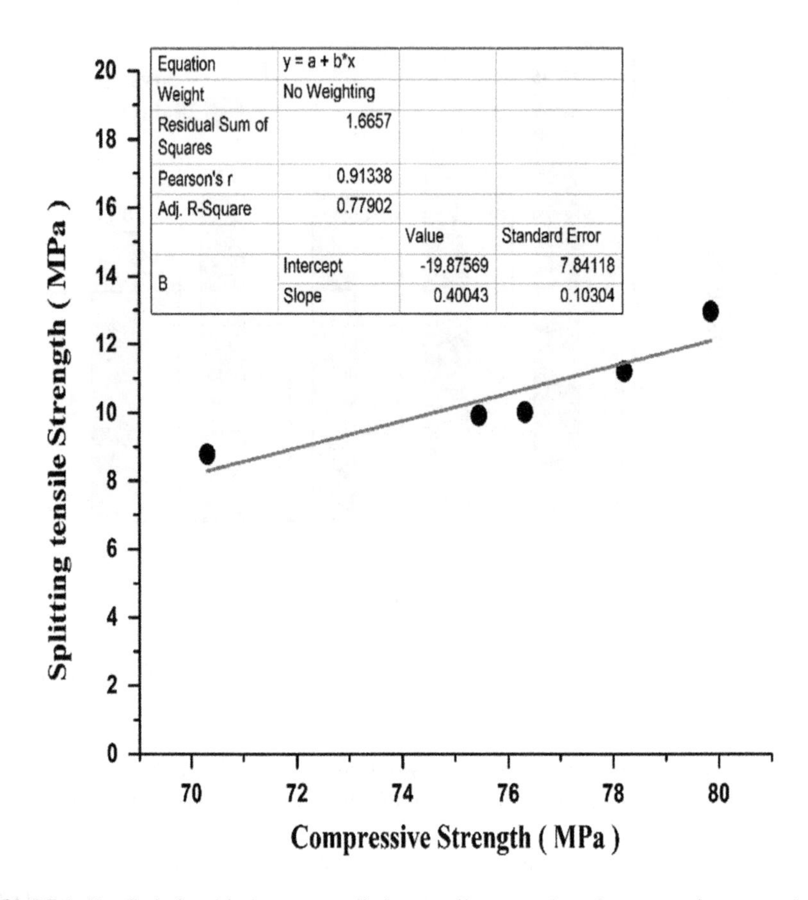

FIGURE 3.12 Relationship between splitting tensile strength and compressive strength.

the range of 8.77 to 12.95 MPa. Although the results show considerable scatter, there is still a tendency for the splitting tensile strength to increase with the compressive strength. The correlation coefficient was found to be 0.91338; this is an indication of a strong positive relationship. Although, there is no direct proportionality in the relation, the ratio of the two depends on the quality of concrete, the aggregate characteristics (surface texture and mineralogy), and the binder quality (Yehia et al., 2015).

Flexural Strength

The flexural strength of a material is defined as its capacity to oppose the bending force applied on the concrete or other slabs placed on the ground. The determination of flexural strength is a prerequisite for the design of concrete mixtures in order to comply with the established standards of an engineering structure. In the present work, room temperature flexural strength was measured according to the specifications of ASTM D790. Figure 3.13 depicts the flexural strength of the prepared F1, F2, F3, F4, and F5 SCGPC specimens, which were discerned to be 6.30, 6.22, 4.99, 4.69, and 4.53 MPa after curing for 7 days, and 7.65, 6.88, 6.12, 5.66, and 4.98 MPa

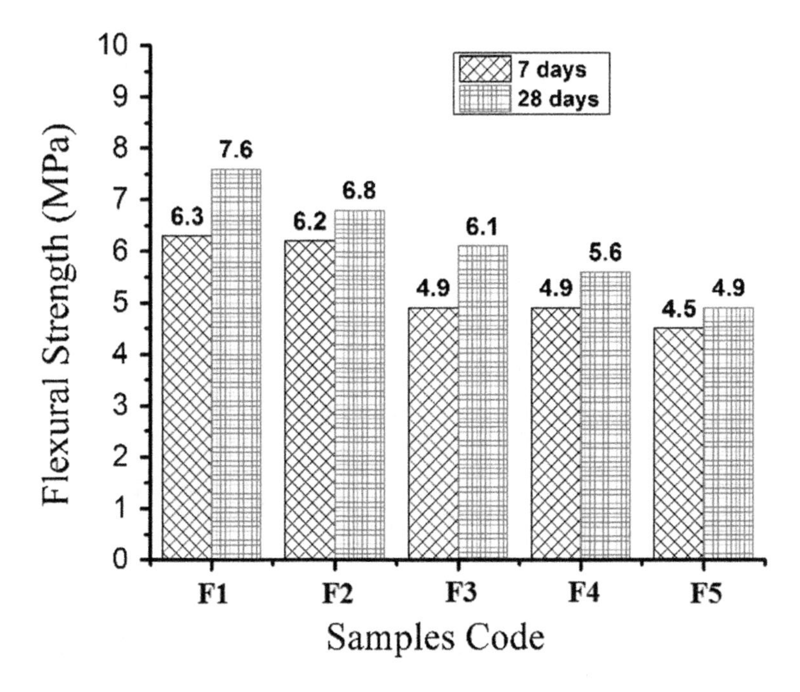

FIGURE 3.13 Curing time-dependent flexural strength of various SCGPC specimens.

after 28 days, respectively. Furthermore, the flexural strength of the spent garnet-incorporated geopolymer mixes in both cases was found to be lower than F1 (the control specimen), which was attributed to the reduction of the sand ratio, which led to the weakening of the bonds between the fine aggregates and binder (Kou, 2009). Table 3.11 shows the raw data for the flexural strengths of various SCGPC specimens.

The regression line approach was used for the relationship between flexural strength and compressive strength. The coefficients of the linear regression equation and the correlation coefficient for the entire data from the specimens are shown in Figure 3.14. Statistical analysis of the results shows that there is a strong relation between the flexural and compressive strengths. It can be seen

TABLE 3.11
Raw Data for Flexural Strength of Various SCGPCs

Sample		7 days (MPa)			28 days (MPa)		
0% garnet	F1	6.83	5.97	6.10	7.42	7.64	7.89
25% garnet	F2	5.93	6.41	6.32	6.45	7.27	6.92
50% garnet	F3	5.01	4.72	5.24	6.01	5.91	6.44
75% garnet	F4	4.70	5.86	4.32	5.22	5.93	5.83
100% garnet	F5	4.72	4.11	4.76	4.86	5.01	5.10

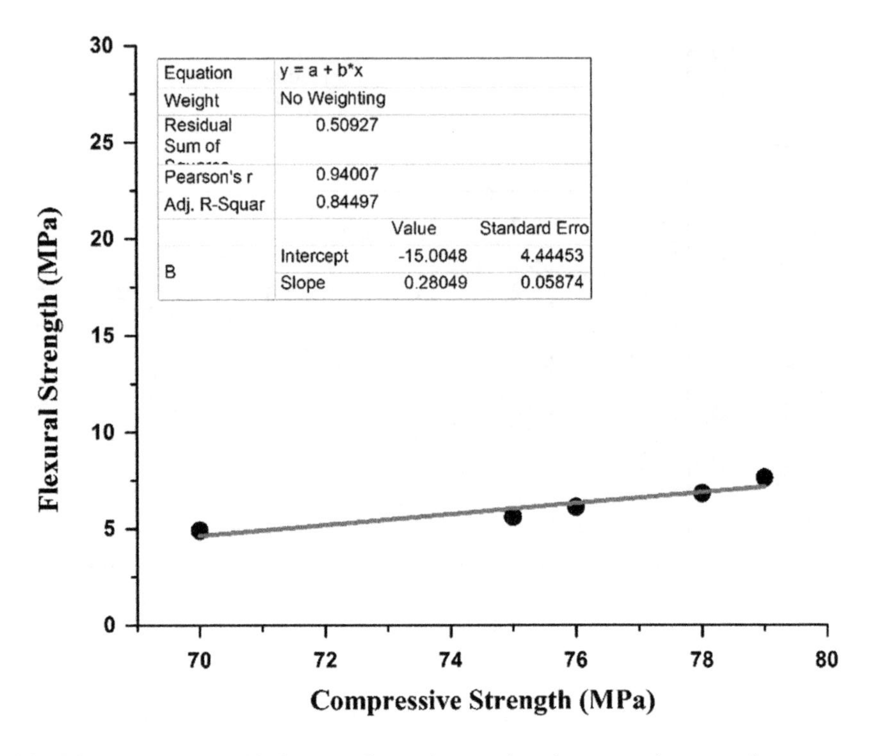

FIGURE 3.14 Relationship between flexural strength and compressive strength.

from the figure that the correlation coefficient for the flexural and compressive strengths is 0.940; this is an indication of a strong positive relationship. The figure also shows that when compressive strength is higher, the flexural strength is also higher.

The relationship between the splitting tensile strength and flexural strength of the SCGPC is shown in Figure 3.15. The correlation was determined for splitting tensile strength in the range of 12.95 to 8.77 MPa and splitting tensile strength in the range of 7.9 to 4.9 MPa. Although the results show considerable scatter, there is still a tendency for the flexural strength to increase as the splitting tensile strength increases. The correlation coefficient was found to be 0.979; this is an indication of a strong positive relationship.

Static Modulus of Elasticity

Mixtures M1 and M2 were prepared to evaluate the Young's modulus (MoE). Table 3.12 lists the MoE values of the M1 and M2 specimens. The SCGPC specimen M2 with 25% spent garnet revealed the highest level of mechanical strength compared with the control throughout the curing period. M2 was selected for this test. The proportions of the M1 and M2 mixes covered the compressive strength in the range of 50 to 80 MPa. The MoE of the SCGPC was evaluated as

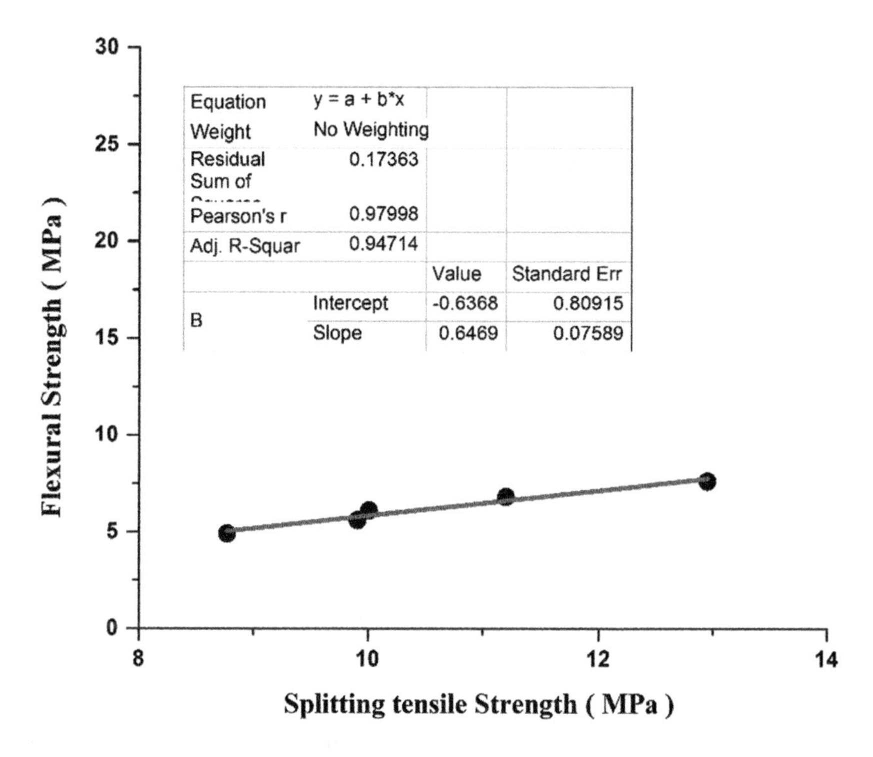

FIGURE 3.15 Relationship between flexural strength and splitting tensile strength.

the secant modulus at a stress level of 40% of the average compressive strength of the concrete cylinders. A total of five concrete cylinders of dimensions 100 ×200 mm were prepared as control and spent garnet mixes, wherein three of them were employed to evaluate the MoE. The average compressive strength was evaluated using the other two cylinders. The MoE ratio was highest for the specimen with maximum compressive strength. The spent garnet reduced the compressive strength of the SCGPC, as mentioned earlier (Table 3.9). This was due to the stronger interfacial bonding between the fine aggregate and bulk matrix of the sand samples. Neville (2011) reported that the MoE of concrete is affected by the aggregate type.

TABLE 3.12
Young's Modulus for TR0 and TR1 Mixes

Concrete Mixes	Mean Compressive Strength (MPa)	Curing Period (days)	MoE Raw Data (GPa)			Average MoE (GPa)
M1	79	28	28.70	32.10	31.6	30.8
M2	78	28	26.30	30.22	27.55	28.3

SUMMARY OF THE CHARACTERIZATION AND THE FRESH AND HARDENED PROPERTIES OF SCGPC

Based on the characterizations of SCGPC constituents such as spent garnet and river sand, and the fresh and hardened state properties of SCGPC, the following conclusions can be drawn.

1. The spent garnet material was found to be well graded, as its grading falls within the upper and lower limits of the medium grading limits (M) for fine aggregate specified in BS 882 (1992). Spent garnet has a large proportion of reactive and inert materials in crystalline form, as revealed by the XRD patterns. The chemical composition of the spent garnet revealed a high silica content as well as iron and alumina. It was noticed that the spent garnet decreased the strength of the SCGPC when it fully replaced sand. The leaching behavior of spent garnet and its safety properties are within the acceptable limits.
2. The performance of spent garnet-based SCGPC was evaluated. A control sample (M1) was designed with a binder-to-solution ratio of 0.4 and a NaOH-to-Na_2SiO_3 ratio of 1:2.5. Subsequently, 25%, 50%, 75%, and 100% of the spent garnet was used to replace the river sand.
3. The SCGPC specimens achieved the desired workability with the inclusion of 3% super-plasticizer and 12% extra water.
4. The mechanical-strength of SCGPC was enhanced with curing age and reduced with increased spent garnet content. It is established that the spent garnet is a prospective candidate for sand replacement in terms of environmental protection and the conservation of natural resources.
5. The substitution of spent garnet in the SCGPC mixes for river sand has seldom been explored. For the first time, the present study has produced spent garnet-based SCGPC with enhanced attributes. Needless to say, the use of spent garnet in concrete as a replacement for traditional river sand has never been explored. The use of spent garnet materials as fine aggregates increased the workability of the concrete. In general, strength of the SCGPC increased with age and decreased with an increase in spent garnet. The results obtained from this chapter demonstrate that it is possible to apply spent garnet in SCGPC.

4 Durability Performance of Self-Compacting Geopolymer Concrete

DURABILITY PROPERTIES OF SELF-COMPACTING GEOPOLYMER CONCRETE

WATER ABSORPTION CAPACITY

Table 4.1 presents the results of a water absorption test on self-compacting geopolymer concrete (SCGPC) conducted with a liquid-to-binder ratio of 0.4 at the curing ages of 7, 28, and 90 days. The permeability of the prepared SCGPC was decreased slightly with increasing age. Furthermore, the control specimen (M1) absorbed less water throughout the curing period compared with all other SCGPC specimens. This could be attributed to the high-water absorbability of the garnet particles (6%) compared with that of the sand particles (3%). Meanwhile, with decreasing curing duration, the water absorption of the concrete specimens decreased. At 7 days, the water absorbed by the M1, M2, M3, M4, and M5 specimens was 3.86%, 4.05% 4.67%, 5.16%, and 5.47%, respectively. Compared with 7 days curing, the water absorption values at 90 days for M1, M2, M3, M4, and M5 were 3.34%, 3.42%, 4.10%, 4.72, and 5.01%, respectively. In fact, there was a reduction by 13%, 15%, 12%, 8.6%, and 8% at 90 days for M1, M2, M3, M4, and M5, respectively, as shown in Figure 4.1. Generally, for all the samples, the water absorption was lower than the limit of 10% recommended by Neville (2011). Chan and Sun (2013) acknowledged that higher water absorption could increase the concrete's permeability.

DRYING SHRINKAGE

Figure 4.2 illustrates the curing time-dependent drying shrinkage of SCGPC specimens. The drying shrinkage of spent garnet-based SCGPC was observed to be lower than SCGPC containing no spent garnet. For the entire duration of the test, the drying shrinkage of the control sample (D1) was superior compared with the samples composed of spent garnet. The drying shrinkage values for D2, D3, D4, and D5 at 90 days were 24%, 49%, 74%, and 86%, respectively, which were lower than the M1 specimen. Table 4.2 shows the drying shrinkage data. Figure 4.2 shows that the drying shrinkage dropped after 35 days, which could be attributed to the rate of moisture loss at an early age in geopolymer concrete; several factors, such as relative humidity, temperature, alkaline solution-to-binder ratio, and the size and shape of the specimen, are thought to affect the shrinkage behavior of geopolymer concrete (Part et al., 2015). The lower drying shrinkage values of the spent garnet-based

TABLE 4.1
Curing Time-Dependent Water Absorption of SCGPC

Samples		Water Absorption (%) Curing Age (days) 7	28	90
0% garnet	M1	3.86	3.62	3.43
25% garnet	M2	4.05	3.87	3.42
50% garnet	M3	4.67	4.36	4.10
75% garnet	M4	5.16	5.06	4.72
100% garnet	M5	5.47	5.22	5.01

SCGPC were attributed to the nature of the fine garnet particles that fill the micropores of the samples and optimize the pore structures. However, the spent garnet increased the water absorption for the samples due to the high-water absorbability of the fine garnet particles (6%) compared with that of sand (3%). This observation agreed with the findings of Zhou (2012). Other factors in the occurrence of drying shrinkage reduction in garnet-mixed concretes may be related to the self-cementing properties of the finer garnet particles. Similar findings revealed that the replacement of river sand in concrete with iron ore tailings could lower shrinkage (Feng, 2011).

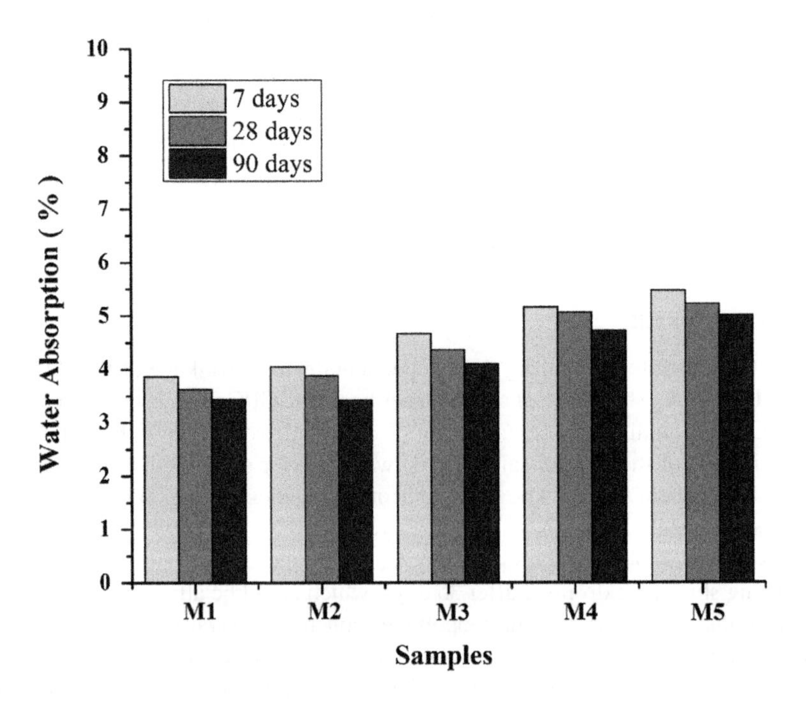

FIGURE 4.1 Water absorption of SCGPC.

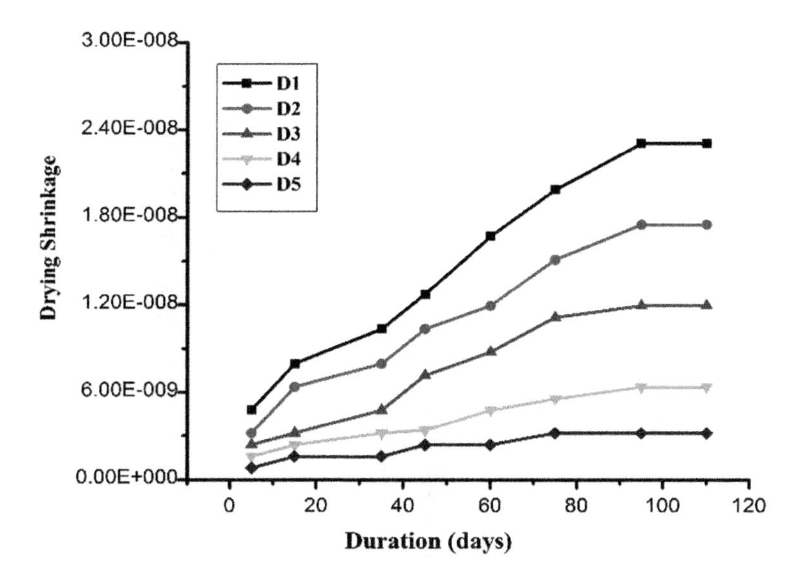

FIGURE 4.2 Curing time-dependent dry shrinkage of SCGPC specimens.

TABLE 4.2
Dry Shrinkage of SCGPC Specimens

	Dry Shrinkage				
Curing Age (days)	**D1 Control**	**D2 25% Garnet**	**D3 50% Garnet**	**D4 75% Garnet**	**D5 100% Garnet**
5	4.776E-9	3.184E-9	2.388E-9	1.592E-9	7.96 E-10
15	7.960E-9	6.368E-9	3.184E-9	2.388E-9	1.592E-9
35	1.034E-8	7.960E-9	4.776E-9	3.184E-9	1.592E-9
45	1.273E-8	1.034E-8	7.164E-9	3.398E-9	2.388E-9
60	1.671E-8	1.194E-8	8.769E-9	4.776E-9	2.388E-9
75	1.990E-8	1.512E-8	1. 114E-8	5.572E-9	3.184E-9
95	2.308E-8	1.751E-8	1.194E-8	6.368E-9	3.184E-9

This was further supported in the study by Zhang (2013) on conventional concrete containing high volumes of iron ore tailings as river sand replacement. It was concluded that the drying shrinkage of concrete made from iron ore tailings was lower than river sand-based concrete.

Resistance to Acid Attacks

Figure 4.3 shows the weight loss of the prepared SCGPC specimens after exposure to acid. Table 4.3 shows the weights of the prepared SCGPC specimens. All the SCGPC specimens revealed a similar decreasing trend in mass for the entire

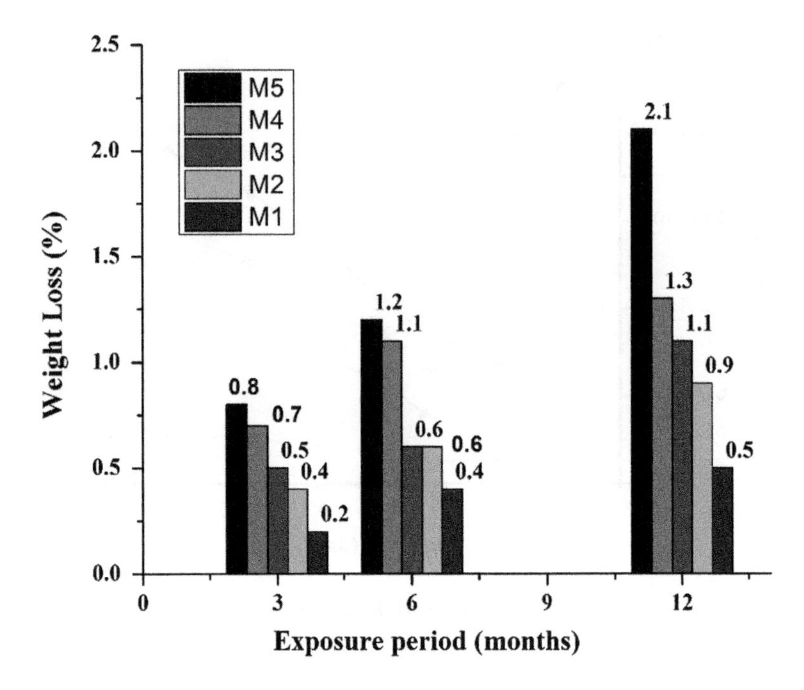

FIGURE 4.3 Weight loss of SCGPC specimens after sulfuric acid exposure.

duration of immersion (3, 6, and 12 months). Throughout the duration of exposure, the weight of the control samples (M1) reduced less than other specimens containing spent garnet. SCGPC with spent garnet revealed very little alteration in the overall look after immersion in H_2SO_4 for 12 months. Figure 4.4 shows the appearance of various concretes upon exposure to H_2SO_4.

There was no change in color after the H_2SO_4 attack. Meanwhile, the garnet samples exhibited the highest weight loss of 2.1%. Conversely, SCGPC specimen M1 manifested the least weight loss (0.5%) in the same 12-month period. The low weight loss of the control samples was attributed to their weak acid absorption tendencies

TABLE 4.3
Weight of the Prepared SCGPC Specimens

Curing Age (Months)	Weight of the Prepared SCGPC Specimens (Kg)									
	M1		M2		M3		M4		M5	
	W1	W2	W1	W2	W1	W2	W1	W2	W1	W2
3	2.320	2.315	2.340	2.330	2.409	2.396	2.445	2.427	2.475	2.455
6	2.318	2.308	2.336	2.321	2.415	2.400	2.440	2.413	2.480	2.450
12	2.323	2.311	2.339	2.317	2.411	2.384	2.442	2.410	2.473	2.421

*W1: Average weight for three samples before sulfuric acid exposure.
*W2: Average weight for three samples after sulfuric acid exposure.

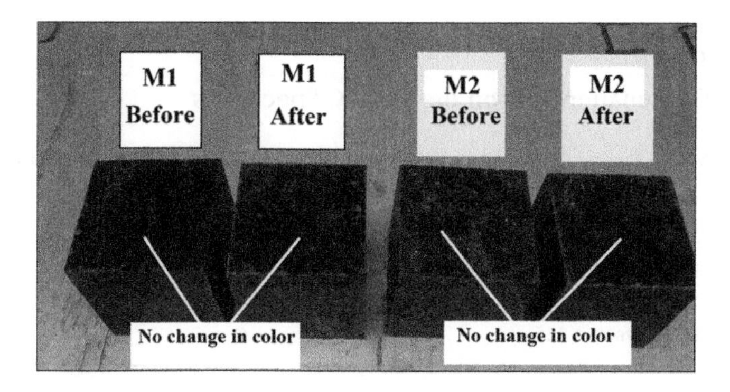

FIGURE 4.4 Appearance of SCGPC concretes upon exposure to H_2SO_4.

as well as the difference in the chemical and phase compositions. Moreover, spent garnet materials that contain low amounts of calcium, as observed in the chemical composition, exhibited weak reactions with sulfuric acid. Singh (2016) also reported similar results on H_2SO_4 exposure for concrete mixes prepared from high levels of low-calcium coal bottom ash (Table 4.3).

Table 4.4 presents the development of the compressive strengths of SCGPC mixes immersed in H_2SO_4 solution. The control samples (M1) demonstrated the best performance, with a mean value of compressive strength reduction of 6.5% after 12 months. This outcome was attributed to the fine particles and the high absorbability of spent garnet, which led to higher acid absorption ratios for the spent garnet samples.

Accelerated Carbonation Depth

Table 4.5 shows the raw data for the carbonation tests on the SCGPC specimens (M1, M2, M3, M4, and M5) containing different percentages of spent garnet. Figures 4.5 and 4.6 show the results for mean carbonation depth, which were reduced steadily

TABLE 4.4
Compressive Strength of SCGPC Mixes Exposed to H_2SO_4 Solution for Different Durations

		Compressive Strength (MPa)			
		Immersed Age (Months)			
Mix		0	3	6	12
0% garnet	M1	79.8	77.2	76.8	74.9
25% garnet	M2	78.2	76.5	75.1	72.6
50% garnet	M3	76.3	74.3	73.6	71.9
75% garnet	M4	75.4	73.8	72.7	70.1
100% garnet	M5	70.3	68.2	67.3	64.4

TABLE 4.5
Carbonation Test on the SCGPC Specimens

Mix		Sample 1 (mm)	Sample 2 (mm)	Average (mm)
0% garnet	M1	13.4	12.2	12.8
25% garnet	M2	11.9	10.5	11.2
50% garnet	M3	8.7	10.3	9.5
75% garnet	M4	9.0	7.6	8.3
100% garnet	M5	5.8	7.0	6.4

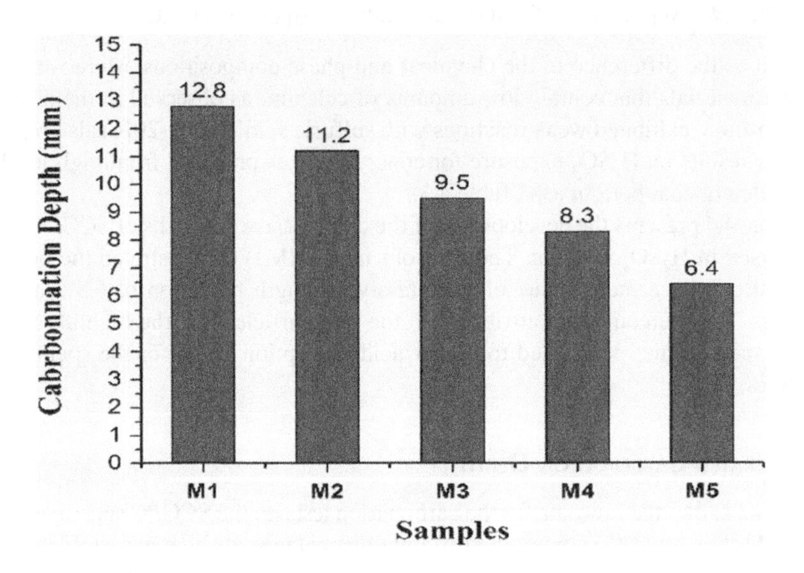

FIGURE 4.5 Carbonation depth of prepared SCGPC.

FIGURE 4.6 Effect of carbonation on the prepared SCGPC specimens.

with increasing percentages of spent garnet in the SCGPC samples. Thus, the prepared SCGPC revealed better performance throughout the replacement of river sand with spent garnet (all percentages). The carbonation rate is usually decided by the concrete's moisture content, the relative humidity (RH) in a given environment, the temperature, and the size of the specimen. One of the foremost factors that influence the concrete's carbonation process is the material's permeability related to its porosity. Therefore, the higher the total porosity, the larger the carbonation depth. Conversely, the carbonation process is slower when the porous structures are finer. Evaluations of some of the durability properties of concretes containing other waste as aggregates displayed no significant effects in carbonation other than that which occurs with natural aggregates (Table 4.5).

Evangelista and De Brito (2010) showed that the carbonation rate of concrete lowered due to the incorporation of fine recycled aggregate in the concrete. This finding agreed with that on waste materials reported by Siddique (2011). It was interpreted in terms of non-interconnected pores that facilitate the ingress of CO_2. According to Basheer (1999), the carbonation rate of concrete is primarily influenced by the tortuosity of the porous network together with the chemistry of the binding phases, water–cement ratio, porosity, and CO_2 transport.

Sulfate Resistance

Figure 4.7 displays the visual look of the prepared SCGPC mixes after immersion in sulfate (Na_2SO_4) solution for the duration (3, 6, and 12 months), which revealed a significant alteration in the overall appearance. The presence of silicate ions in the activation solution led to the development of more compacted structures with rich silica gels. The SCGPC specimens all revealed similar decreasing trends in mass for the entire duration of immersion (3, 6, and 12 months). Throughout the duration of exposure, the weight of the control samples (M1) was reduced less than that of other specimens containing spent garnet.

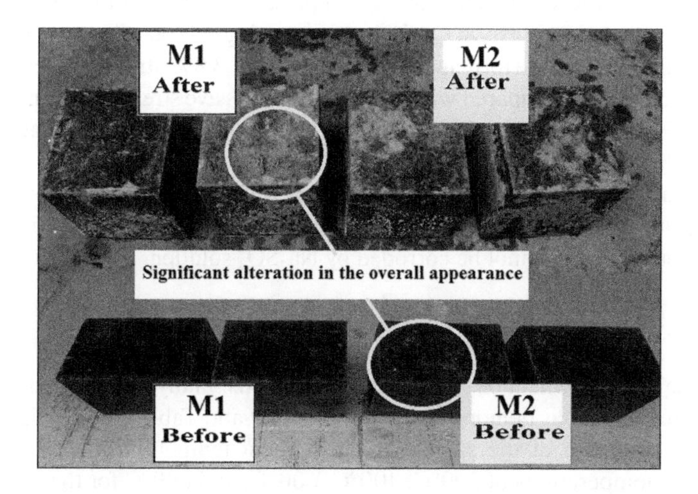

FIGURE 4.7 Appearance of various SCGPC mixes subjected to sulfate attack.

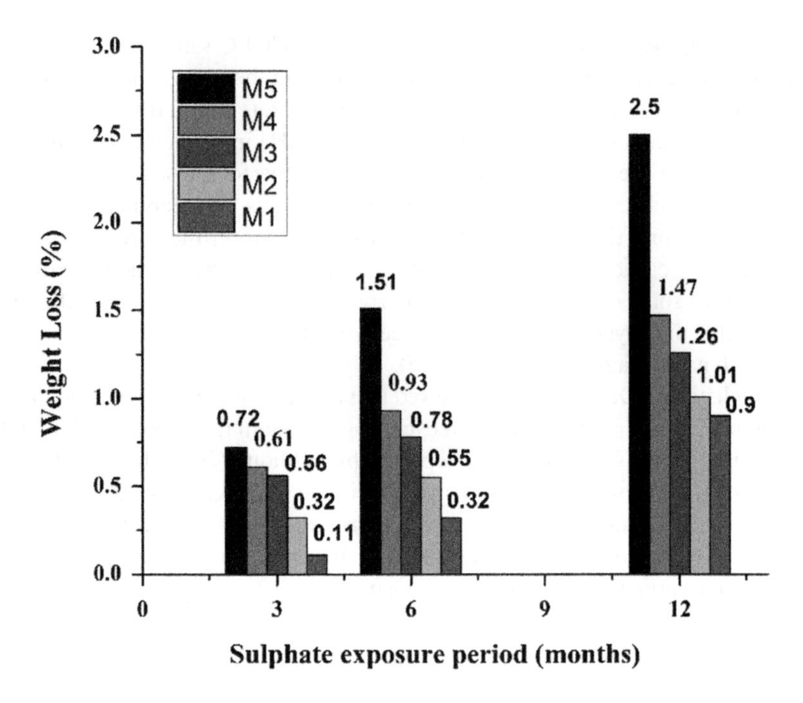

FIGURE 4.8 Sulfate exposure duration-dependent weight loss of SCGPC.

Figure 4.8 illustrates the weight loss of the respective SCGPC mixes depending on the duration of exposure to Na_2SO_4 solution. The garnet samples (M5) exhibited the highest weight loss at 2.5%. Conversely, SCGPC control specimens TR0 manifested the lowest weight loss of 0.9% in the same 12-month period. The lower weight loss of the control samples was attributed to their weak (Na_2SO_4) absorption tendencies as well as the differences in their chemical and phase compositions.

Table 4.6 lists the development of the compressive strengths of the SCGPC specimens after immersion in Na_2SO_4 solution for varying periods. The control samples (M1) display the best performance with an average 3.8% strength decrease (from 79.84 to 76.28 MPa), compared with the 7.4% compressive strength declination (from 70.32 to 65.71 MPa) of the garnet (M5) specimen. In general, geopolymer products do not contain $Ca(OH)_2$ due to being formed from source materials containing little or no calcium. As such, when these materials are immersed in Na_2SO_4 solution, there is no gypsum or ettringite growth in the matrix to cause expansion, meaning that geopolymer concrete cannot be corroded by Na_2SO_4 solution.

Performance under Elevated Temperatures

During the heat treatment of the concrete specimens, many surface and color variations were evident, as demonstrated in Figure 4.9 and Table 4.7. The figure shows the initial condition of the specimens at ambient temperature (28°C) and after exposure to temperatures of 200°C, 400°C, 600°C, and 800°C for the control (M1) and spent garnet SCGPC (M2). It was found that the color of the M1 and M2 mixes

TABLE 4.6
Compressive Strengths of SCGPC Mixes after Subjection to Sulfate Attack for Different Durations

	Compressive Strength (MPa)			
	Immersed Age (Months)			
Mixes	**0**	**3**	**6**	**12**
M1	79	78	77	76
M2	78	77	75	74
M3	76	75	74	72
M4	75	73	71	69
M5	70	69	66	65

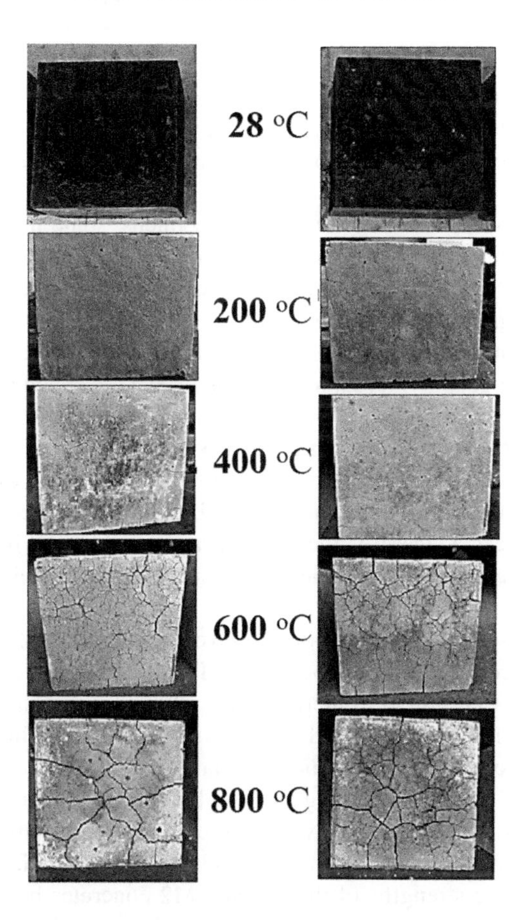

FIGURE 4.9 Typical look of the surface texture of the control (M1) and the SCGPC (M2) mixes exposed to various elevated temperatures.

TABLE 4.7
Physical Properties of SCGPC Mixes Subjected to Various Elevated Temperatures

Temperature (°C)	Specimen	Shape	Color	Texture
28	Control (M1)	Perfect cube (PC)	Black	Smooth
	Garnet (M2)	PC	Black	Smooth
200	Control (M1)	PC	Light gray	Smooth
	Garnet (M2)	PC	Light gray	Smooth
400	Control (M1)	PC	Light gray	Smooth
	Garnet (M2)	PC	Light gray	Smooth
600	Control (M1)	Rough edge	Light gray	Crack/depression
	Garnet (M2)	Rough edge	Light gray	Crack/depression
800	Control (M1)	Distorted	Whitish gray	Crack/depression
	Garnet (M2)	Distorted	Whitish gray	Crack/depression

at ambient temperature was black with a smooth surface and perfect edges. These properties were retained up to 28°C. However, at 200°C both the control specimen and spent garnet concretes developed incipient cracks with perfect edges and a light gray color.

At a higher temperature (600°C) for all specimens, some hairline cracks were observed that were light gray in color and with rough edges. At an even higher temperature (800°C), both M1 and M2 mixes developed wide cracks over the surface that were whitish gray in appearance.

During the heating process, some transformations may occur that lead to an acceleration in the geopolymerization process of SCGPC, such as moisture evaporation, chemical decomposition, and internal vapor pressure. In the premature heating period (at low temperatures), the lack of transformations might prevent the development of cracks. However, at higher temperatures, the moisture content of the concrete was lost and the plastic limit was exceeded, resulting in the appearance of cracks. Generally, the color variations observed in the M1 and M2 concretes could be related to the chemical changes in the specimen occurring at higher temperatures.

Figure 4.10 presents the residual compressive strengths of the SCGPC specimens subjected to different elevated temperatures. The initial compressive strengths of all the specimens were found to increase up to 200°C, which may be ascribed to the further hydration of unhydrated slag grains due to steaming. For spent garnet concrete that contains pozzolanic material, an additional calcium aluminum silicate hydrate (C-A-S-H) gel was generated because of the pozzolanic reaction. Savva et al. (2005) reported that elevated temperatures had a similar effect on the mechanical attributes of concretes made from limestone and silica aggregates. The residual compressive strengths of the M1 and M2 concretes begin to decrease at 400°C. The observed reduction in compressive strength with the increase of temperature is attributed to the dehydration of the C-A-S-H gel system at about 500°C.

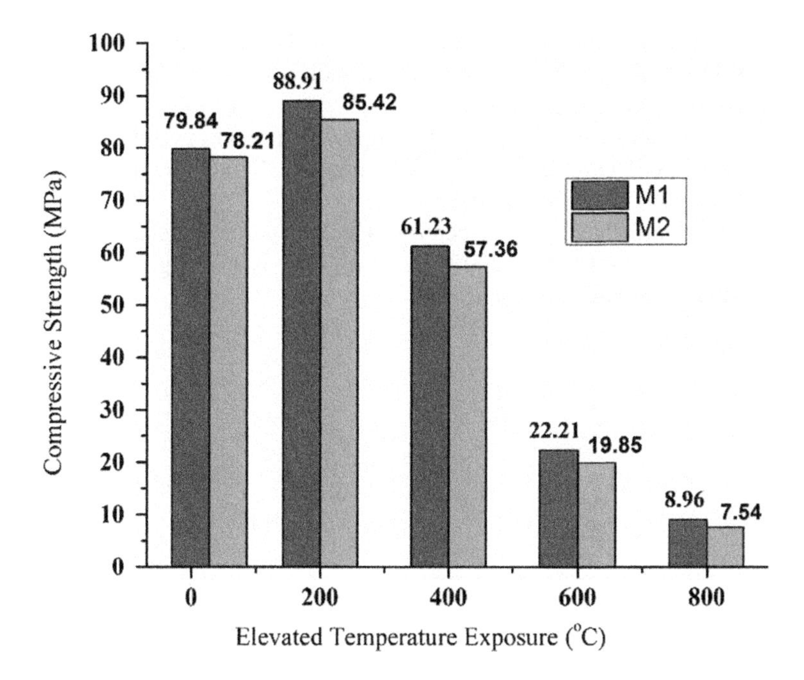

FIGURE 4.10 Residual compressive strengths of SCGPC specimens subjected to different elevated temperatures.

Over 600°C, strength losses are mainly caused by calcium carbonate dissociation and the subsequent escape of CO_2 from $CaCO_3$. Based on the outcome, three temperature ranges were demarcated: (1) 28°C–200°C, where strength gains were generally observed for all the concrete specimens, (2) 200°C–400°C, where the strength falls were relatively stiff, and (3) 400°C–800°C, where the strength values reduced significantly.

SUMMARY

This chapter has examined the durability of control and spent garnet-based SCGPCs exposed to various deleterious agents such as elevated temperatures, carbonation, sulfates, and acid attacks. Overall, the study has revealed that spent garnet-based concrete offers commendable attributes in terms of resistance to aggressive environments. Some of the other advantages based on various results can be emphasized as follows:

1. The water absorption of spent garnet-based SCGPC was (5.01%) slightly higher than that of the control (3.43%). Moreover, all the test results were below the specified limits.
2. The drying shrinkage of the spent garnet-based SCGPC was lower than that of concrete made with river sand.
3. Spent garnet-based SCGPC demonstrated poorer performance with respect to acid attacks.

4. The mean carbonation depth of spent garnet-based SCGPC mixes (6.4%) was lower than the control (12.8%), indicating a better performance throughout the replacement of river sand.
5. Spent garnet-based SCGPC demonstrated poorer performance with respect to sulfate attacks. The garnet-samples exhibited the highest weight loss at 2.5%.
6. The thermal behavior of the control as well as garnet-based SCGPC samples under exposure up to 200°C displayed increased strength and a change in color from black to light gray.

5 Results on Morphology, Bonding, and Thermal Properties

MORPHOLOGY ANALYSIS

The morphology of self-compacting geopolymer concrete (SCGPC) including spent garnet was investigated using various characterization techniques. Two SCGPC specimens with 0% (control) and 25% (optimum) spent garnet were selected to determine their microstructures after being exposed to acid for 6 months. As mentioned in the previous chapter, the SCGPC specimen with 25% spent garnet revealed the highest level of mechanical strength compared with the control throughout the curing period. The results obtained from the microstructure characterizations are discussed and analyzed in the upcoming sections. SCGPC specimens were characterized using field emission scanning electron microscopy (FESEM), X-ray diffusion (XRD), and Fourier-transform infrared (FTIR) measurements. Furthermore, the dehydration (mass loss) and the decomposition of the basic hydration products were determined by means of thermogravimetric analysis (TGA) and differential thermal analysis (DTA).

FIELD EMISSION SCANNING ELECTRON MICROSCOPY IMAGES

FESEM imaging was performed to acquire useful information regarding the presence of spent garnet in the SCGPC and to distinguish the morphological features responsible for strength enhancement. The microstructure of the SCGPC consisted of geopolymer paste, aggregates, and an interfacial transitional zone. Figures 5.1 and 5.2 show the FESEM images of the control (M1) and the SCGPC mix (M2) after 6 months of curing, respectively.

FESEM analysis clearly revealed a reduction in porosity due to the addition of slag to the SCGPC mix, where the matrix appeared compacted with improved space-filling features through the slag-activated creation of calcium alumina silicate hydrate (C-A-S-H) gel. The solid matrix developed good mechanical properties. The FESEM micrograph of the SCGPC specimens demonstrated an alteration in morphology due to the creation of bonds among the geopolymer paste and the aggregate. The formation of such bonds allowed the nucleation of new randomly oriented crystal structures on a small scale (Figure 5.2). Eventually, large crystalline grains were spread over the entire specimen volume. This in turn generated cracks in the garnet geopolymer paste and reduced the mechanical strength, as seen in the compressive strength values of the M2 sample, which was lower than the purely river sand-based

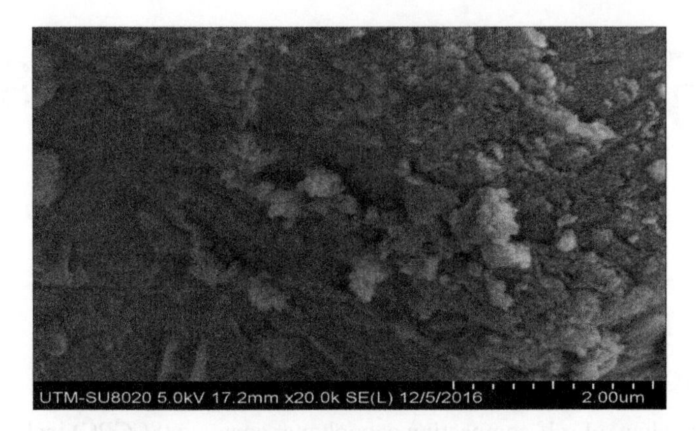

FIGURE 5.1 FESEM images of SCGPC specimen C-A-S-H gel M1 (2 μm scale).

FIGURE 5.2 FESEM images of SCGPC specimen M2 (2 μm scale).

SCGPC specimen (M1). It was demonstrated that the presence of spent garnet in the mix remarkably affected the surface morphology and the morphological properties of the SCGPC. The particle shapes of the spent garnet in M2 are shown in Figure 5.2.

X-Ray Diffraction Pattern

The XRD analysis provided useful information on the structural properties (crystalline phases and lattice orientations) of the hardened concretes. Figure 5.3 illustrates the XRD patterns of the SCGPC specimens M1 (control) and M2 (containing 25% spent garnet) at the age of 6 months. The diffractograms revealed the occurrence of hydrated phases, including quartz (SiO_2), hematite (Fe_2O_3), albite ($NaAlSi_3O_8$), and nepheline ($3Na_2OK_2O4Al_2O_39Si$), with a peak of C-A-S-H ($CaAl_2Si_2O_8$). However, M2 contains calcium (CaO) in addition to the chemical elements listed in M1.

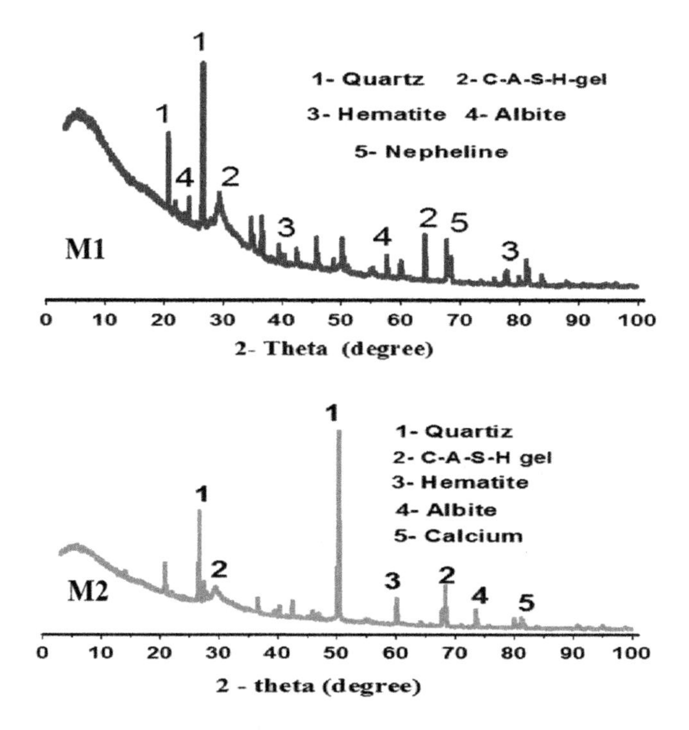

FIGURE 5.3 XRD patterns of specimens M1 and M2.

The primary phases in the SCGPC samples were found to be quartz, albite, and C-A-S-H. Besides, the phases in the concrete, including quartz and C-A-S-H showed similar XRD patterns in the 2θ range of 20°C to 25°C. The detected calcium and nepheline phases originated from the geopolymerization reaction (Ohje, 2011). Meanwhile, no ettringite phase was detected in the proposed SCGPC. The C-A-S-H peak area for M1 appeared higher than that of M2; this led to lower strength in the M2 samples. This in turn reduced the mechanical strength, as seen in the compressive strength values of the M2 sample, which was lower than the purely river sand-based SCGPC specimen (M1).

BONDING AND THERMAL ANALYSIS

Bonding Analysis

Figures 5.4 and 5.5 depict the FTIR spectra of the SCGPC specimens M1 (control) and M2 (containing 25% spent garnet) at the age of 6 months. The chemical bonding vibrations (i.e., molecular characteristics) in the proposed SCGPC specimens were evaluated in terms of IR absorption or transmission properties using FTIR spectroscopy. The observed IR broad band at around 3847 to 3835 cm^{-1} characterizes the O-H stretching vibration and H-O-H bending vibration, respectively. These modes of vibration are due to the weakly bound water molecules that were adsorbed on the surface or trapped in the large cavities. Both M1 and M2 specimens revealed

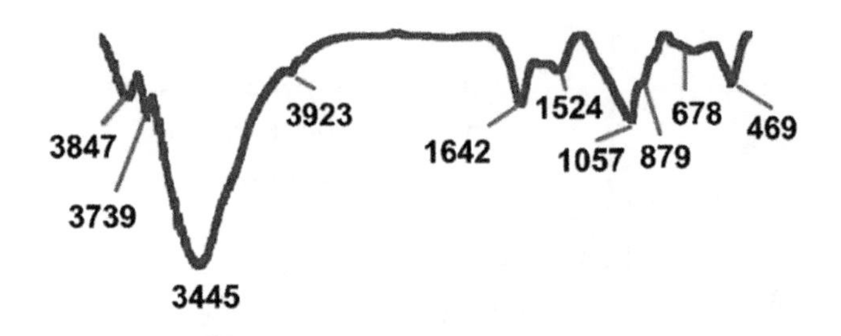

FIGURE 5.4 FTIR spectra of the control SCGPC sample M1.

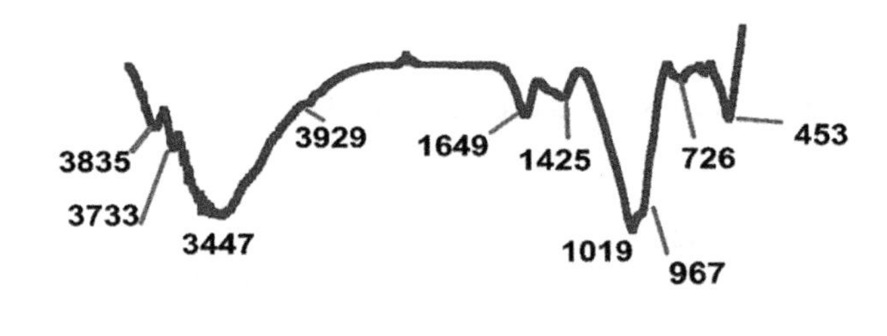

FIGURE 5.5 FTIR spectra of the spent garnet-based SCGPC sample M2.

primarily similar spectra corresponding to the hydroxyl group in the range of 469 to 3847 cm^{-1} and 453 to 3835 cm^{-1}, respectively. The appearance of a prominent band in the range of 1019 to 1057 cm^{-1} and a relatively weak band at 1425 cm^{-1} were attributed to the presence of sodium carbonate (Na_2CO_3); this corresponds to the stretching vibration of the O-C-O bond (Swanepoel, 2002). The band corresponding to Si-O-Si/Si-O-Al bending was evidenced at 879 to 967 cm^{-1}, which verified the bonding of alumina in the proposed SCGPC. Such bands are commonly observed in ring silicate materials due to their non-crystallization signature, because the intensity

TABLE 5.1
Functional Groups of Bonding Analysis

Wavenumbers (cm^{-1})	Bond	Standard
3847–3835	H-O-H	*PR
1019–1057	O-C-O	*PR
879–967	Si-O-Si/Si-O-Al	*PR
1624–1649	O-H	*PR

*PR: Refers to previous research of standard method.

of these bands is independent on the extent of crystallization. The appearance of broad bands in the wavenumber range of 1624 to 1649 cm^{-1} were explained by the stretching of the O-H and H-O-H groups, and the deformation vibrations originated from the weakly bound water molecules. Such water molecules were adsorbed on the surface of the specimens or trapped within the outsized cavities between the rings of the three-dimensional network. The IR broad bands at 3847 to 3835 cm^{-1} of the control and spent garnet-based concretes were attributed to the O-H and H-O-H groups' stretching vibrational modes (Fernández-Jiménez, 2005a,b). Due to the presence of water molecules, these bands reduced the transmittance intensity of the spent garnet-based SCGPC (Table 5.1).

Thermal Analysis

Figures 5.6 and 5.7 present the TGA/DTA curves of SCGPC specimens M1 and M2, respectively, after subjection to 6 months of curing. The TGA/DTA analyses in the temperature range of 30°C to 1000°C were performed to evaluate the thermal stability of the SCGPC prepared with spent garnet. The temperature domains corresponding to the thermal decomposition of the different phases of the concretes were identified. SCGPC samples M1 (without spent garnet) and M2 (with 25% spent garnet) were selected for thermal analysis (for reasons explained previously). It was found that both samples revealed mass losses or gains due to thermal decomposition, oxidation, and hydration. The temperature range in which the mass losses of the three identified phases of the respective concretes corresponded to 40°C–160°C (phase 1), 550°C–650°C (phase 2), and 750°C–830°C (phase 3). Phase 1 exhibited great amounts of mass loss due to the rapid evaporation of adsorbed free water (Bondar, 2011). Subsequently, water was also lost from the carbo-aluminate hydrates part. Similar observations were made by Nikolov and Rostovsky (2014). Phase 2 begun with the dehydration of Na_2CO_3, which means that a considerable portion of the activator did not take part in the activation reaction. Phase 3 originated from the dehydroxlation of the binder paste (slag) and the phase transition of quartz. This phase revealed an endothermic peak (Morsya et al., 2008). Nonetheless, the DTA curve of the M1 and M2 specimens indicated similar phase transitions to TGA in three confined phases. In phase 1,

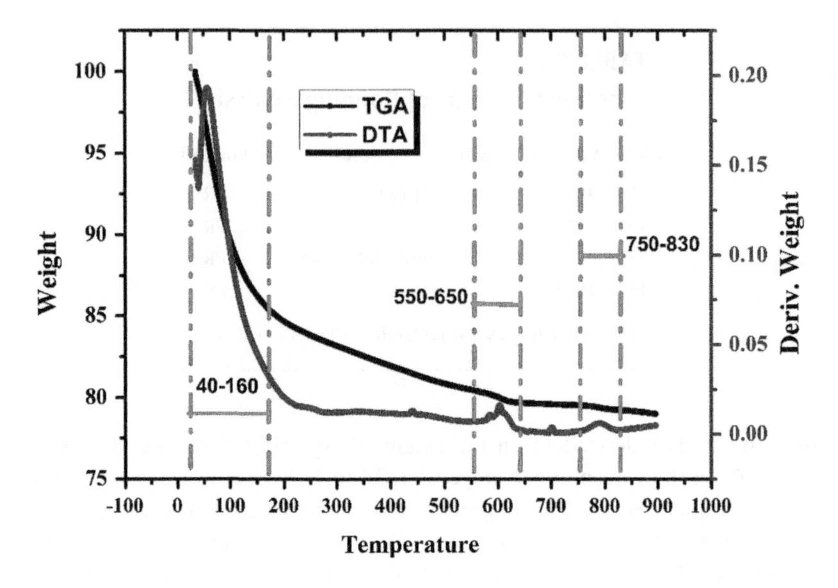

FIGURE 5.6 TGA/DTA curves for SCGPC sample M1.

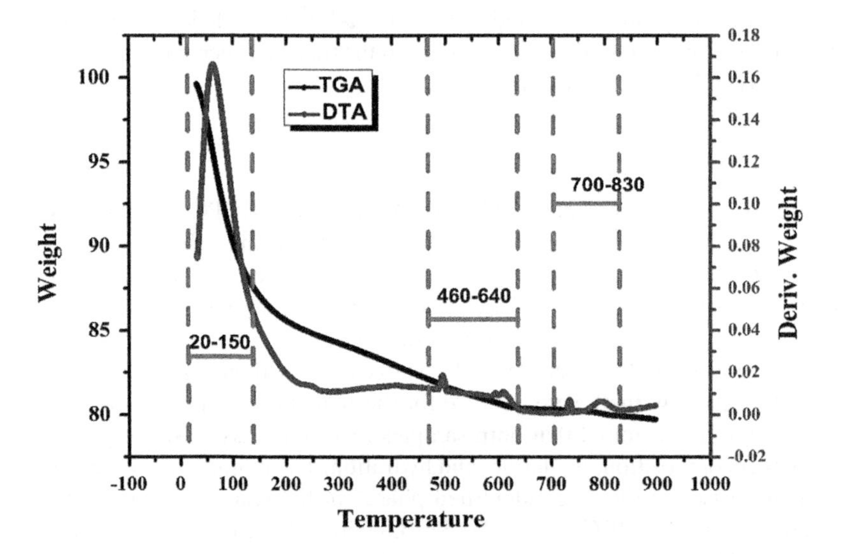

FIGURE 5.7 TGA/DTA curves for SCGPC sample M2.

the dehydration reaction of the carbo-aluminate produced C-A-S-H. This phase formation is mainly attributable to the reactivity of microfine garnet particles, which can generate secondary C-A-S-H, thereby contributing to the enhancement of compressive strength. Besides, these hydration products were responsible for the densification and enhancement of the mechanical, deformation, and durability properties of the SCGPC.

SUMMARY

1. This chapter has provided the results on the microstructures of the prepared SCGPC specimens. The FESEM micrograph of the specimens demonstrated an alteration in morphology due to the formation of bonds between the GP paste and the aggregate. This bonding allowed the nucleation of new randomly oriented crystal structures on a small scale (Figure 5.1b). Eventually, large crystalline grains were spread over the entire specimen volume. This in turn generated cracks in the garnet GP paste and reduced the mechanical strength, as seen in the compressive strength values of the M2 sample, which was lower than the purely river sand-based SCGPC specimen (M1).
2. The XRD pattern showed the presence of hydrated phases such as quartz (SiO_2), albite (NaAlSi3O8), and nepheline (3Na2OK2O4Al2O39Si) with a peak of C-A-S-H. The C-A-S-H peak area for the M1 control sample was higher than that of the M2 spent garnet sample, which led to lower strength in the M2 sample. This in turn reduced the mechanical strength, as seen in the compressive strength values of the M2 sample.
3. The chemical bonding vibrations (i.e., molecular characteristics) in the proposed SCGPC specimens were evaluated in terms of IR absorption or transmission properties using FTIR spectroscopy. The IR broad band observed from 3847 to 3835 cm^{-1} characterizes the O-H stretching vibrations and H-O-H bending vibrations, respectively. Both specimens TR0 and TR1 revealed primarily similar spectra corresponding to the hydroxyl group in the ranges of 469 to 3847 cm^{-1} and 453 to 3835 cm^{-1}, respectively.
4. The TGA/DTA analyses in the temperature range of 30°C to 1000°C were performed to evaluate the thermal stability of the SCGPC prepared with spent garnet. It was found that both samples M1 and M2 revealed mass loss or gain due to thermal decomposition, oxidation, and hydration.

6 Overall Performance of Spent Garnet as Sand Replacement in Self-Compacting Geopolymer Concrete

CHARACTERIZATION OF CONSTITUENT MATERIALS

Spent garnet, having a large proportion of reactive and inert materials, can form crystalline phases in the presence of irregular structures with high specific surface areas, as revealed in the X-ray diffusion (XRD) patterns. The chemical composition of spent garnet reveals high silica, alumina, and iron contents. The leaching behavior of spent garnet is within the acceptable limits. Moreover, spent garnet material is recommended as a well-graded fine aggregate appropriate for self-compacting geopolymer concrete (SCGPC) manufacturing. The fineness modulus of spent garnet is slightly lower than the recommended minimum value.

FRESH CONCRETE PROPERTIES

The fresh characteristics of SCGPC made with spent garnet in place of river sand revealed satisfactory performance and elevated compressive strength. Slag as a geopolymer binder phase can be synergistically combined with conventional sand, spent garnet, and coarse aggregates to enhance SCGPC. A spent garnet replacement level of 100% gave the optimum performance in terms of flow ability.

HARDENED CONCRETE PROPERTIES

The hardened state density of the spent garnet samples was higher in the control samples for SCGPC. The compressive, flexural, and tensile strengths of the spent garnet-based SCGPC were lower than the control specimen. Thus, spent garnet was demonstrated to have potential for sand replacement up to 100%. The Young's modulus (MoE) of SCGPC containing spent garnet was comparable to that of the control throughout the testing period. The drying shrinkage performance of the spent garnet-based SCGPC was better than the control specimen.

DURABILITY

Experimental results on water absorption in the early stages (7 days) reveal that it was higher for the SCGPC specimens containing spent garnet than in the control specimen. But as the duration of curing was increased to 90 days, the absorption rate decreased in all the SCGPC specimens. In short, the water absorption of spent garnet-based SCGPC was within acceptable limits at less than 10%. The performance of spent garnet-based SCGPC exposed to carbon dioxide environments showed excellent resistance to rapid carbon dioxide penetration. A negative mass change of the spent garnet-based SCGPC specimen was observed due to acid exposure. SCGPC using spent garnet as replacement for river sand demonstrated lower performance against sulfate attacks.

MORPHOLOGY ANALYSIS

The morphology analyses of the constituent materials of the concrete mix and the SCGPC including spent garnet as replacement for river sand provided invaluable information about concrete at the micro scale. It constitutes a complimentary basis to the traditional methods of evaluating microstructures. Field emission scanning electron microscopy (FESEM) of the SCGPC demonstrated an alteration in morphology due to the creation of bonds among the geopolymer paste and the fine aggregates of spent garnet. This bonding in turn allowed the nucleation of new randomly oriented crystal structures on a small scale. Eventually, large crystalline grains spread over the entire volume of the specimen. This in turn generated cracks in the garnet geopolymer paste and reduced the mechanical strength. The XRD pattern showed the appearance of hydrated phases, including SiO_2, Fe_2O_3, $NaAlSi_3O_8$, and $3Na_2OK_2O_4Al_2O_{39}Si$ with a peak of calcium alumina silicate hydrate (C-A-S-H). Briefly, the intensity of the XRD peaks decreased when river sand was replaced with spent garnet, whereas the typical C-A-S-H phase was increased with hydration.

BONDING AND THERMAL ANALYSIS

The chemical bonding vibrations (i.e., molecular characteristics) in the proposed SCGPC specimens were evaluated in terms of IR absorption or transmission properties using Fourier-transform infrared (FTIR) spectroscopy. The observed IR broad band at around 3847 to 3835 cm^{-1} characterizes the O-H stretching vibration and H-O-H bending vibration, respectively. These modes of vibration are due to the weakly bound water molecules that were adsorbed on the surface or trapped in large cavities. Both the TR0 and TR1 specimens revealed primarily similar spectra corresponding to hydroxyl groups in the range of 469 to 3847 cm^{-1} and 453 to 3835 cm^{-1}, respectively. TGA/DTA analyses in the temperature range of 30°C to 1000°C were performed to evaluate the thermal stability of the SCGPC prepared with spent garnet. It was found that both samples TR0 and TR1 revealed mass loss or gain due to thermal decomposition, oxidation, and hydration.

BOOK CONTRIBUTION

For the first time, the present study has created spent garnet-based SCGPC with enhanced attributes. Needless to say, the use of spent garnet in concrete as a replacement for traditional river sand has seldom been explored. Thus, none or very limited information was available in the literature on the long-term effects, especially the corrosiveness of the material. The use of spent garnet materials such as recycled waste products, ceramic waste, and coal bottom ash as aggregates in construction materials has been rapidly evolving from a concept to laboratory studies to field work worldwide. The long-term impact of the inclusion of spent garnet in concretes is still at the development stage. The results obtained from this study demonstrate that it is possible to use spent garnet in SCGPC without compromising the expected long-term quality. Although the various applications of spent garnet have yet to be exploited, it is expected that the realization of these benefits will not only promote the sustainability of the concrete-making industry but will also have a long-standing impact on promoting the implementation of waste materials for civil engineering.

Appendix: Raw Data

Sieve Analysis: Garnet Attributes

Sieve Size (mm)	Weight Retained (g)	Cumulative Weight Retained (g)	Weight Passed (g)	Passing (%)	Retained (%)
9.5	0	0	500	100	0
5.0	0	0	500	100	0
2.36	11	11	489	97.8	2.2
1.18	14.5	25.5	474.5	94.9	5.1
0.6	122	147.5	352 .5	70.5	29.5
0.3	228.5	376	124	24.8	75.2
0.15	103	479	21	6.2	93.8
PAN	20.5	500	0	0	0
Fineness modulus = cumulative retained (%) / 100				2.05	

Sieve Analysis: Sand Attributes

Sieve Size (mm)	Weight Retained (g)	Cumulative Weight Retained (g)	Weight Passed (g)	Passing (%)	Retained (%)
9.5	0	0	500	100	0
5.0	12	12	488	98	2
2.36	57	69	431	86	14
1.18	80	149	351	70	30
0.6	107	256	244	49	51
0.3	128	384	116	23	77
0.15	78	462	38	8	92
PAN	38	500	0	0	0
Fineness modulus = cumulative retained (%) / 100				2.6	

Specimen		Sample 1	Sample 2	Sample 3	Average
		Compressive Strength (MPa): 3 Days			
0% garnet	C1-3	69.84	71.01	69.75	70.24
25% garnet	C2-3	71.96	70.96	72.55	71.62
50% garnet	C3-3	67.13	66.80	68.63	67.50
75% garnet	C4-3	65.11	67.04	65.63	65.91
100% garnet	C5-3	60.69	61.93	61.64	61.42
		Compressive Strength (MPa): 7 Days			
0% garnet	C1-7	73.55	72.98	73.34	73.29
25% garnet	C2-7	72.23	74.45	73.07	73.13
50% garnet	C3-7	70.19	69.97	71.10	70.42
75% garnet	C4-7	69.10	71.03	67.41	69.18
100% garnet	C5-7	66.69	67.13	64.77	66.23
		Compressive Strength (MPa): 28 Days			
0% garnet	C1-28	80.09	78.15	81.28	79.84
25% garnet	C2-28	77.53	79.63	77.47	78.21
50% garnet	C3-28	74.16	76.87	77.93	76.32
75% garnet	C4-28	73.85	76.15	75.40	75.45
100% garnet	C5-28	68.96	70.22	71.78	70.32

Splitting Tensile Strength (MPa): 7 Days

Specimen		Sample 1	Sample 2	Sample 3	Average
0% garnet	T1-7	10.30	9.70	10.40	10.10
25% garnet	T2-7	10.35	9.81	9.60	9.92
50% garnet	T3-7	9.52	8.84	7.93	8.76
75% garnet	T4-7	7.79	7.91	8.65	8.10
100% garnet	T5-7	7.40	7.25	7.97	7.54

Splitting Tensile Strength (MPa): 28 Days

Specimen		Sample 1	Sample 2	Sample 3	Average
0% garnet	T1-28	12.55	12.41	13.90	12.95
25% garnet	T2-28	10.95	11.11	11.42	11.20
50% garnet	T3-28	9.65	10.02	10.36	10.01
75% garnet	T4-28	9.51	10.22	10.01	9.91
100% garnet	T5-28	8.36	9.01	8.94	8.77

Flexural Strength (MPa): 7 Days					
Specimen		**Sample 1**	**Sample 2**	**Sample 3**	**Average**
0% garnet	F1-7	6.83	5.97	6.10	6.30
25% garnet	F2-7	5.93	6.41	6.32	6.22
50% garnet	F3-7	5.01	4.72	5.24	4.99
75% garnet	F4-7	4.70	5.86	4.32	4.96
100% garnet	F5-7	4.72	4.11	4.76	4.53

Flexural Strength (MPa): 28 Days					
Specimen		**Sample 1**	**Sample 2**	**Sample 3**	**Average**
0% garnet	F1-28	7.42	7.64	7.89	7.65
25% garnet	F2-28	6.45	7.27	6.92	6.88
50% garnet	F3-28	6.01	5.91	6.44	6.12
75% garnet	F4-28	5.22	5.93	5.83	5.66
100% garnet	F5-28	4.86	5.01	5.10	4.99

Mix		Mean Compressive Strength (MPa)	Curing Period (days)	MoE Raw Data (GPa)			Average MoE (GPa)
				Sample 1	Sample 2	Sample 3	Average
0% garnet	M1	79	28	28.70	32.10	31.6	30.8
25% garnet	M2	78	28	26.30	30.22	27.55	28.3

Fresh Properties of the Prepared SCGPC					
Mix		**Slump Flow (mm)**	**T50 cm Slump Flow (seconds)**	**V-Funnel (seconds)**	**L-Box Ratio (H_2/H_1)**
0% garnet	M1	671	5.5	12	0.91
25% garnet	M2	675	5	11.5	0.92
50% garnet	M3	681	4.5	11	0.93
75% garnet	M4	692	4	7.5	0.95
100% garnet	M5	700	3.5	6.5	0.97

Mix		Sample 1	Sample 2	Sample 3	Average
		Water Absorption (%): 7 Days			
0% garnet	M1	3.70	3.96	3.92	3.86
25% garnet	M2	4.07	3.98	4.10	4.05
50% garnet	M3	4.81	4.25	4.95	4.67
75% garnet	M4	5.03	4.96	5.49	5.16
100% garnet	M5	5.40	5.15	5.86	5.47
		Water Absorption (%): 28 Days			
0% garnet	M1	3.95	3.20	3.71	3.62
25% garnet	M2	3.85	4.09	3.67	3.87
50% garnet	M3	4.41	4.57	4.10	4.36
75% garnet	M4	4.93	5.16	5.09	5.06
100% garnet	M5	5.15	5.02	5.49	5.22
		Water Absorption (%): 90 Days			
0% garnet	M1	3.23	3.62	3.44	3.43
25% garnet	M2	3.57	3.48	3.21	3.42
50% garnet	M3	4.19	3.97	4.14	4.10
75% garnet	M4	4.81	4.45	4.90	4.72
100% garnet	M5	5.10	4.93	5.0	5.01

Curing Age (Months)	Weight Loss of the Prepared SCGPC Specimens after Sulfate Exposure (kg)														
	M1 Control			M2			M3			M4			M5		
	W1	W2	W%	W1	W2	W%	W1	W2	W%	W1	W2	W%	W1	W2	W%
3	2.295	2.291	0.11	2.333	2.325	0.32	2.401	2.387	0.56	2.432	2.417	0.61	2.465	2.447	0.72
6	2.301	2.293	0.32	2.327	2.314	0.55	2.407	2.388	0.78	2.445	2.422	0.93	2.472	2.434	1.51
12	2.320	2.299	0.90	2.330	2.306	1.01	2.416	2.385	1.26	2.438	2.402	1.47	2.460	2.398	2.5

*W1: Average weight of three samples before sulfate exposure.

*W2: Average weight of three samples after sulfate exposure.

Compressive Strength of SCGPC Mixes Exposed to H_2SO_4 Solution

Specimen		Sample 1	Sample 2	Sample 3	Average
		Compressive Strength (MPa): 3 Months			
0% garnet	C1-3	77.41	78.05	76.30	77.25
25% garnet	C2-3	76.43	77.12	76.10	76.55
50% garnet	C3-3	72.98	74.82	75.22	74.34
75% garnet	C4-3	75.10	72.36	74.03	73.83
100% garnet	C5-3	66.03	70.13	68.53	68.23
		Compressive Strength (MPa): 6 Months			
0% garnet	C1-6	76.55	77.75	76.10	76.80
25% garnet	C2-6	73.46	74.86	77.01	75.11
50% garnet	C3-6	72.61	74.28	74.02	73.63
75% garnet	C4-6	73.82	71.91	72.60	72.78
100% garnet	C5-6	68.57	65.73	67.62	67.31
		Compressive Strength (MPa): 12 Months			
0% garnet	C1-12	75.24	73.72	75.83	74.93
25% garnet	C2-12	72.33	74.11	71.39	72.61
50% garnet	C3-12	71.89	73.06	70.80	71.92
75% garnet	C4-12	68.66	71.15	70.49	70.10
100% garnet	C5-12	62.69	64.50	66.07	64.42

Weight Loss of the Prepared SCGPC Specimens after Sulfuric Acid Exposure (kg)

Curing age (Months)	M1 control			M2			M3			M4			M5		
	W1	W2	W%	W1	W2	W%	W1	W2	W%	W1	W2	W%	W1	W2	W%
3	2.320	2.315	0.2	2.340	2.330	0.4	2.409	2.396	0.5	2.445	2.427	0.7	2.475	2.455	0.8
6	2.318	2.308	0.4	2.336	2.321	0.6	2.415	2.400	0.6	2.440	2.413	1.1	2.480	2.450	1.2
12	2.323	2.311	0.5	2.339	2.317	0.9	2.411	2.384	1.1	2.442	2.410	1.3	2.473	2.421	2.1

*W1: Average weight of three samples before sulfuric acid exposure.

*W2: Average weight of three samples after sulfuric acid exposure.

Compressive Strength of SCGPC Mixes after Subjection to Sulfate Attack

Specimen		Sample 1	Sample 2	Sample 3	Average
		Compressive Strength (MPa): 3 Months			
0% garnet	C1-3	78.21	79.05	77.10	78.12
25% garnet	C2-3	75.60	77.13	78.57	77.10
50% garnet	C3-3	74.63	75.43	76.86	75.64
75% garnet	C4-3	72.98	75.01	71.70	73.23
100% garnet	C5-3	67.08	71.09	69.13	69.10
		Compressive Strength (MPa): 6 Months			
0% garnet	C1-6	76.10	78.06	77.87	77.34
25% garnet	C2-6	75.05	76.93	75.71	75.90
50% garnet	C3-6	72.90	75.10	74.66	74.22
75% garnet	C4-6	70.16	73.30	72.39	71.95
100% garnet	C5-6	68.03	65.89	66.14	66.69
		Compressive Strength (MPa): 12 Months			
0% garnet	C1-12	75.89	77.02	75.93	76.28
25% garnet	C2-12	73.91	75.79	74.61	74.77
50% garnet	C3-12	70.28	72.63	73.45	72.12
75% garnet	C4-12	68.76	71.13	69.78	69.89
100% garnet	C5-12	67.39	64.62	65.12	65.71

Carbonation Test on the SCGPC Specimens

Mix	Sample 1 (mm)	Sample 2 (mm)	Average (mm)
M1	13.4	12.2	12.8
M2	11.9	10.5	11.2
M3	8.7	10.3	9.5
M4	9.0	7.6	8.3
M5	5.8	7.0	6.4

Dry Shrinkage of SCGPC Specimens L0–L1

Curing Age (Days)	L0	D1 L1	D2 L1	D3 L1	D4 L1	D5 L1
5	2.917	2.911	2.913	2.914	2.915	2.916
15	2.917	2.907	2.909	2.913	2.914	2.915
35	2.917	2.904	2.907	2.911	2.913	2.915
45	2.917	2.901	2.904	2.908	2.912	2.914
60	2.917	2.896	2.902	2.906	2.911	2.914
75	2.917	2.892	2.898	2.903	2.910	2.913
95	2.917	2.888	2.895	2.902	2.909	2.913

$$\text{Dry shrinkage} = \left(\frac{Lo-L}{100}\right) \times \text{gauge factor.}$$

Curing Age (Days)	D1	D2	D3	D4	D5
5	$\frac{2.917-2.911}{100}\times 0.0000796$	$\frac{2.917-2.913}{100}\times 0.0000796$	$\frac{2.917-2.914}{100}\times 0.0000796$	$\frac{2.917-2.915}{100}\times 0.0000796$	$\frac{2.917-2.916}{100}\times 0.0000796$
	$=4.776E-9$	$=3.184E-9$	$=2.388E-9$	$=1.592E-9$	$=7.96\ E-10$
15	$\frac{2.917-2.907}{100}\times 0.0000796$	$\frac{2.917-2.909}{100}\times 0.0000796$	$\frac{2.917-2.913}{100}\times 0.0000796$	$\frac{2.917-2.914}{100}\times 0.0000796$	$\frac{2.917-2.915}{100}\times 0.0000796$
	$=7.960E-9$	$=6.368E-9$	$=3.184E-9$	$=2.388E-9$	$=1.592E-9$
35	$\frac{2.917-2.904}{100}\times 0.0000796$	$\frac{2.917-2.907}{100}\times 0.0000796$	$\frac{2.917-2.911}{100}\times 0.0000796$	$\frac{2.917-2.913}{100}\times 0.0000796$	$\frac{2.917-2.915}{100}\times 0.0000796$
	$=1.034E-8$	$=7.960E-9$	$=4.776E-9$	$=3.184E-9$	$=1.592E-9$
45	$\frac{2.917-2.901}{100}\times 0.0000796$	$\frac{2.917-2.904}{100}\times 0.0000796$	$\frac{2.917-2.908}{100}\times 0.0000796$	$\frac{2.917-2.912}{100}\times 0.0000796$	$\frac{2.917-2.914}{100}\times 0.0000796$
	$=1.273E-8$	$=1.034E-8$	$=7.164E-9$	$=3.398E-9$	$=2.388E-9$
60	$\frac{2.917-2.896}{100}\times 0.0000796$	$\frac{2.917-2.902}{100}\times 0.0000796$	$\frac{2.917-2.906}{100}\times 0.0000796$	$\frac{2.917-2.911}{100}\times 0.0000796$	$\frac{2.917-2.914}{100}\times 0.0000796$
	$=1.671E-8$	$=1.194E-8$	$=8.769E-9$	$=4.776E-9$	$=2.388E-9$
75	$\frac{2.917-2.892}{100}\times 0.0000796$	$\frac{2.917-2.898}{100}\times 0.0000796$	$\frac{2.917-2.903}{100}\times 0.0000796$	$\frac{2.917-2.910}{100}\times 0.0000796$	$\frac{2.917-2.913}{100}\times 0.0000796$
	$=1.990E-8$	$=1.512E-8$	$=1.\ 114E-8$	$=5.572E-9$	$=3.184E-9$
95	$\frac{2.917-2.888}{100}\times 0.0000796$	$\frac{2.917-2.895}{100}\times 0.0000796$	$\frac{2.917-2.902}{100}\times 0.0000796$	$\frac{2.917-2.909}{100}\times 0.0000796$	$\frac{2.917-2.913}{100}\times 0.0000796$
	$=2.308E-8$	$=1.751E-8$	$=1.194E-8$	$=6.368E-9$	$=3.184E-9$

Fresh Density (kg/m³)

Concrete Mixes	Volume (m³)	Mass 1 (kg)	Mass 2 (kg)	Fresh Density (kg/m³)
M1	0.010	7.8	31.6	2380
M2	0.010	7.8	31.85	2405
M3	0.010	7.8	32.0	2420
M4	0.010	7.8	32.44	2464
M5	0.010	7.8	32.73	2493

Hardened State Density of SCGPC (28 Days)

Concrete Mixes	Volume (m³)	Average Mass (kg)	Density (kg/m³)
M1	0.001	2.370	2370
M2	0.001	2.395	2395
M3	0.001	2.410	2410
M4	0.001	2.445	2445
M5	0.001	2.471	2471

Physical Properties of SCGPC Mixes Subjected to Various Elevated Temperatures

Temperature (°C)	Mix	Shape	Color	Texture
28	Control (M1)	Perfect cube (PC)	Black	Smooth
	Garnet (M2)	PC	Black	Smooth
200	Control (M1)	PC	Light gray	Smooth
	Garnet (M2)	PC	Light gray	Smooth
400	Control (M1)	PC	Light gray	Smooth
	Garnet (M2)	PC	Light gray	Smooth
600	Control (M1)	Rough edge	Light gray	Crack/depression
	Garnet (M2)	Rough edge	Light gray	Crack/depression
800	Control (M1)	Distorted	Whitish gray	Crack/depression
	Garnet (M2)	Distorted	Whitish gray	Crack/depression

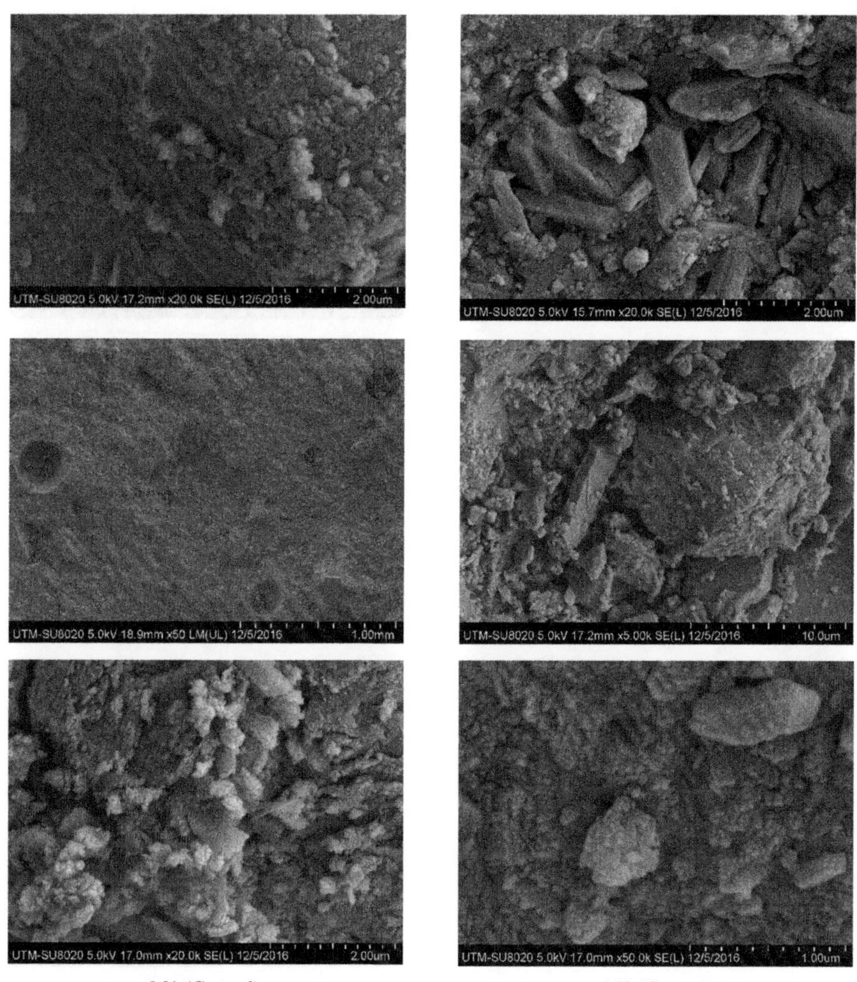

M1 (Control) M2 (Garnet)

FESEM images of SCGPC specimens M1 and M2.

FESEM images of sand and garnet.

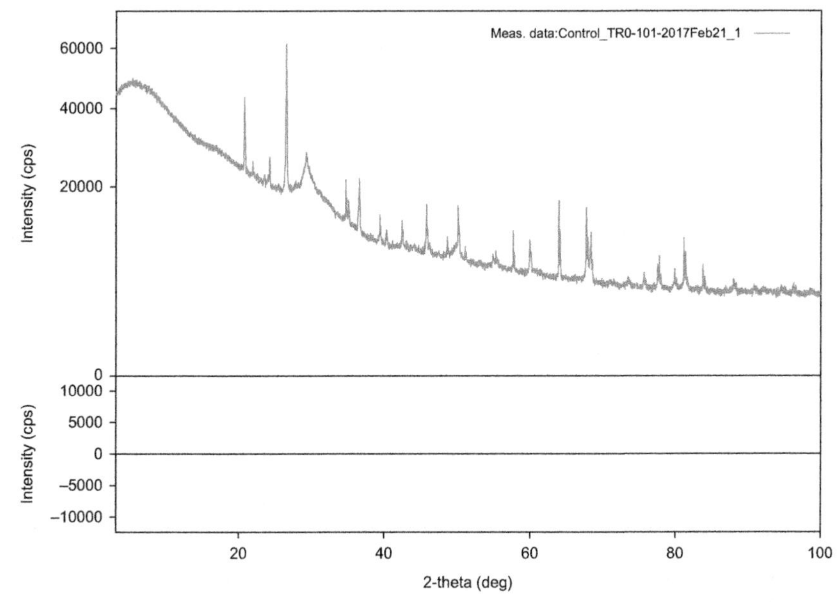

XRD Pattern Control Mix 1

No.	2-theta (deg)	d(ang.)	Height (cps)	FWHM (deg)	Int. I(cps deg)	Int. W(deg)	Asym. factor
1	7.9(4)	11.2(6)	356(50)	1.8(6)	918(200)	2.6(9)	0.5(6)
2	17.2(5)	5.15(14)	324(47)	2.2(4)	755(233)	2.3(11)	1.1(11)
3	20.823(8)	4.2625(16)	12958(299)	0.132(6)	1898(78)	0.146(9)	0.64(16)
4	24.326(9)	3.6560(14)	5206(189)	0.090(15)	699(53)	0.134(15)	1.2(6)
5	26.572(7)	3.3518(9)	11364(280)	0.080(14)	1366(425)	0.12(4)	1.8(16)
6	26.650(3)	3.3422(4)	33600(481)	0.081(9)	3897(459)	0.116(15)	0.9(3)
7	29.29(3)	3.047(3)	3804(162)	1.03(3)	4407(160)	1.16(9)	0.52(8)
8	34.794(4)	2.5763(3)	8637(244)	0.040(8)	563(37)	0.065(6)	1.2(7)
9	35.149(9)	2.5511(7)	3523(156)	0.080(17)	573(40)	0.163(18)	1.1(6)
10	36.488(5)	2.4605(3)	1875(114)	0.06(2)	171(67)	0.09(4)	3(5)
11	36.641(6)	2.4506(4)	8428(241)	0.124(9)	1541(97)	0.183(17)	2.8(8)
12	39.457(7)	2.2819(4)	4022(166)	0.061(12)	470(24)	0.117(11)	1.6(9)
13	40.339(18)	2.2340(10)	1711(108)	0.12(2)	223(33)	0.13(3)	1.1(6)
14	42.486(5)	2.1260(2)	4388(174)	0.070(5)	473(18)	0.108(8)	1.1(3)
15	45.843(8)	1.9778(3)	6821(217)	0.079(12)	1023(48)	0.150(12)	1.3(6)
16	48.689(7)	1.8686(3)	2235(124)	0.063(7)	151(17)	0.068(11)	2.1(11)
17	50.02(8)	1.822(3)	1161(89)	1.08(5)	1346(58)	1.16(14)	1.6(4)
18	50.132(4)	1.81817(14)	6074(204)	0.101(5)	653(49)	0.108(12)	0.51(10)
19	51.088(11)	1.7864(4)	1736(109)	0.080(10)	149(15)	0.086(14)	0.57(14)
20	54.890(18)	1.6713(5)	1220(92)	0.15(3)	312(30)	0.26(4)	0.9(6)
21	55.277(9)	1.6605(2)	1692(108)	0.072(13)	212(22)	0.13(2)	0.3(3)

(Continued)

No.	2-theta (deg)	d(ang.)	Height (cps)	FWHM (deg)	Int. I(cps deg)	Int. W(deg)	Asym. factor
22	57.684(3)	1.59683(7)	5281(191)	0.047(3)	353(16)	0.067(5)	0.9(2)
23	59.984(8)	1.54096(19)	4100(168)	0.145(8)	799(31)	0.195(16)	1.0(3)
24	63.9718(19)	1.45418(4)	12312(291)	0.079(2)	1205(22)	0.098(4)	1.39(15)
25	67.725(3)	1.38243(6)	10638(271)	0.111(3)	1463(34)	0.138(7)	0.97(13)
26	68.048(8)	1.37666(15)	2140(121)	0.34(3)	820(60)	0.38(5)	0.20(6)
27	68.366(5)	1.37103(9)	4164(169)	0.082(7)	423(54)	0.101(17)	0.8(2)
28	75.702(13)	1.25535(18)	1357(97)	0.145(11)	209(20)	0.15(3)	2.0(8)
29	77.593(6)	1.22942(8)	2407(129)	0.103(10)	399(22)	0.166(18)	1.5(5)
30	77.815(3)	1.22646(4)	2416(129)	0.052(6)	201(22)	0.083(13)	1.6(5)
31	79.945(9)	1.19906(11)	2135(121)	0.102(13)	382(18)	0.179(19)	1.4(7)
32	81.197(3)	1.18370(4)	6155(206)	0.064(3)	534(47)	0.087(10)	0.9(2)
33	81.460(17)	1.1805(2)	1395(98)	0.26(4)	505(49)	0.36(6)	2.2(7)
34	83.791(6)	1.15353(7)	2560(133)	0.091(7)	291(17)	0.114(13)	0.52(19)
35	87.964(6)	1.10925(7)	1129(88)	0.177(19)	250(22)	0.22(4)	0.3(2)
36	96.27(3)	1.0343(2)	668(68)	0.14(3)	98(24)	0.15(5)	1.6(15)

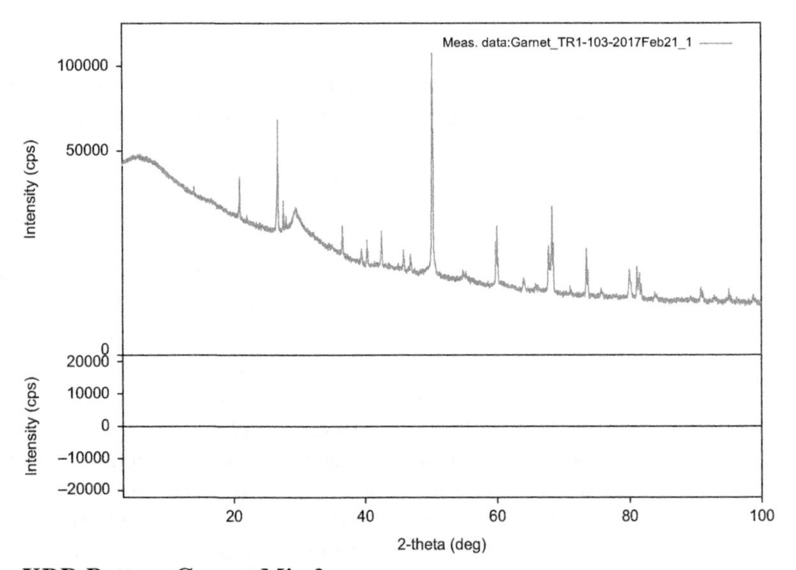

XRD Pattern Garnet Mix 2

No.	2-theta (deg)	d(ang.)	Height (cps)	FWHM (deg)	Int. I(cps deg)	Int. W (deg)	Asym. factor
1	8.1(3)	10.9(4)	476(57)	1.8(3)	899(225)	1.9(7)	0.7(6)
2	20.944(4)	4.2382(8)	13009(299)	0.062(5)	1207(55)	0.093(6)	1.1(4)
3	26.707(3)	3.3352(4)	40093(525)	0.055(5)	4256(103)	0.106(4)	2.1(7)
4	27.561(7)	3.2338(8)	8515(242)	0.036(4)	330(42)	0.039(6)	1.1(8)
5	29.39(4)	3.037(4)	3053(145)	0.99(4)	3221(169)	1.06(11)	0.62(11)
6	36.4(2)	2.469(14)	387(52)	2.0(3)	910(126)	2.3(6)	4(5)
7	36.577(6)	2.4547(4)	6868(217)	0.083(6)	716(80)	0.104(15)	2.1(7)
8	39.490(7)	2.2801(4)	2973(143)	0.079(12)	336(28)	0.113(15)	2.9(17)
9	40.330(5)	2.2345(3)	5282(191)	0.078(6)	501(32)	0.095(9)	2.0(7)
10	42.527(2)	2.12404(11)	8603(243)	0.081(3)	1125(21)	0.131(6)	1.08(15)
11	45.857(5)	1.97722(19)	3811(162)	0.084(7)	445(32)	0.117(13)	0.51(12)
12	46.902(9)	1.9356(4)	3196(148)	0.094(8)	381(26)	0.119(14)	0.7(3)
13	50.2036(16)	1.81576(5)	97460(819)	0.1016(14)	13605(98)	0.140(2)	1.27(9)
14	54.86(2)	1.6721(7)	634(66)	0.87(8)	590(71)	0.9(2)	0.27(17)
15	59.828(3)	1.54462(8)	9235(252)	0.087(6)	1112(64)	0.120(10)	0.70(12)
16	59.976(4)	1.54116(9)	9124(251)	0.092(5)	1161(62)	0.127(10)	0.70(13)
17	64.057(16)	1.4525(3)	1574(104)	0.189(17)	317(29)	0.20(3)	1.9(7)
18	65.84(3)	1.4174(6)	662(67)	0.15(7)	206(32)	0.31(8)	0.4(6)
19	67.728(3)	1.38239(5)	7039(220)	0.137(8)	1298(62)	0.184(15)	0.21(3)
20	67.836(3)	1.38045(6)	9423(255)	0.014(5)	230(35)	0.024(4)	2(2)
21	68.308(3)	1.37205(5)	19664(368)	0.108(4)	3511(60)	0.179(6)	1.13(13)
22	73.495(4)	1.28750(6)	8894(247)	0.115(4)	1315(27)	0.148(7)	0.83(12)
23	75.718(17)	1.2551(2)	1015(84)	0.175(15)	190(21)	0.19(4)	0.9(4)
24	79.988(6)	1.19852(7)	4309(172)	0.215(6)	1181(23)	0.274(16)	1.33(17)
25	81.145(6)	1.18433(7)	5577(196)	0.145(4)	956(31)	0.171(12)	0.84(13)
26	81.560(6)	1.17935(7)	3445(154)	0.120(7)	492(26)	0.143(14)	1.5(3)
27	90.838(12)	1.08149(11)	1934(115)	0.191(10)	406(24)	0.21(3)	1.1(3)
28	95.100(18)	1.04395(15)	1189(90)	0.15(3)	264(23)	0.22(4)	1.2(7)

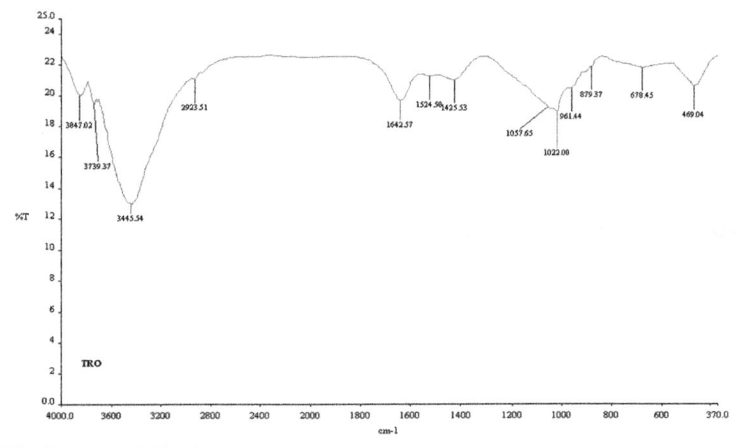

FTIR Control: Mix 1

4000	22.563	3992	22.45355	3984	22.38441	3976	22.27942	3968	22.23694
3999	22.54534	3991	22.45761	3983	22.3803	3975	22.27355	3967	22.22298
3998	22.52778	3990	22.46421	3982	22.38172	3974	22.26964	3966	22.20033
3997	22.51039	3989	22.46412	3981	22.37986	3973	22.26237	3965	22.17448
3996	22.50012	3988	22.45465	3980	22.36775	3972	22.25283	3964	22.15255
3995	22.48995	3987	22.4381	3979	22.34499	3971	22.24569	3963	22.13726
3994	22.47486	3986	22.41773	3978	22.31766	3970	22.24303	3962	22.12427
3993	22.45976	3985	22.3979	3977	22.29388	3969	22.24208	3961	22.10757

3960	22.08776	3952	21.89583	3944	21.73117	3936	21.70067	3928	21.38974
3959	22.07046	3951	21.85143	3943	21.75163	3935	21.60071	3927	21.36255
3958	22.05767	3950	21.82343	3942	21.74743	3934	21.51351	3926	21.35313
3957	22.04624	3949	21.81255	3941	21.72042	3933	21.46386	3925	21.3525
3956	22.03148	3948	21.79976	3940	21.71663	3932	21.45715	3924	21.33345
3955	22.01074	3947	21.77032	3939	21.75501	3931	21.46587	3923	21.29171
3954	21.98182	3946	21.73587	3938	21.79189	3930	21.45675	3922	21.25838
3953	21.94273	3945	21.71975	3937	21.77586	3929	21.42626	3921	21.24398

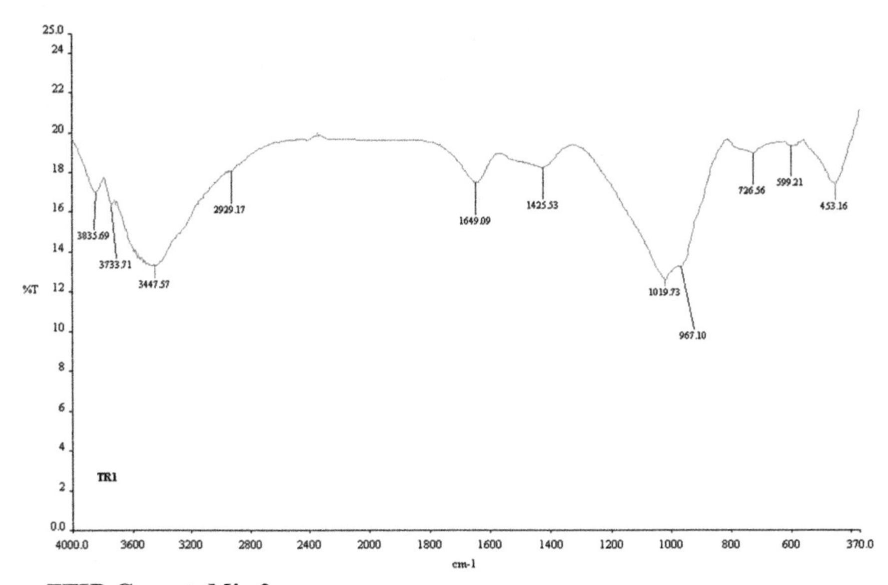

FTIR Garnet: Mix 2

4000	19.66816	3982	19.48433	3964	19.23726	3942	18.83239	3924	18.41657
3999	19.65205	3981	19.47848	3963	19.22158	3941	18.80271	3923	18.37496
3998	19.63606	3980	19.46284	3962	19.20725	3940	18.79878	3922	18.34383
3997	19.62022	3979	19.43852	3961	19.18938	3939	18.83479	3921	18.33106
3996	19.60954	3978	19.41202	3960	19.16932	3938	18.86555	3920	18.32419
3995	19.59684	3977	19.39045	3959	19.15206	3937	18.84414	3919	18.33402
3994	19.57868	3976	19.37764	3958	19.13879	3936	18.7697	3918	18.37594
3993	19.56214	3975	19.37121	3957	19.12625	3935	18.67787	3917	18.42559
3992	19.55611	3974	19.36485	3956	19.11028	3934	18.60235	3916	18.44517
3991	19.56071	3973	19.35494	3955	19.08866	3933	18.56214	3915	18.44178
3990	19.56726	3972	19.34382	3954	19.05954	3932	18.55772	3914	18.44772
3989	19.56687	3971	19.33587	3953	19.02138	3931	18.56143	3913	18.46729
3988	19.55755	3970	19.3322	3952	18.9769	3930	18.54487	3912	18.47158
3987	19.54173	3969	19.32952	3951	18.93582	3929	18.50855	3911	18.43946
3986	19.52245	3968	19.3223	3950	18.91026	3928	18.46945	3910	18.3685
3985	19.50375	3967	19.3068	3949	18.89941	3927	18.44328	3909	18.26957
3984	19.49083	3966	19.28376	3948	18.88505	3926	18.4367	3908	18.15553
3983	19.48579	3965	19.25855	3947	18.85507	3925	18.43704	3907	18.04357

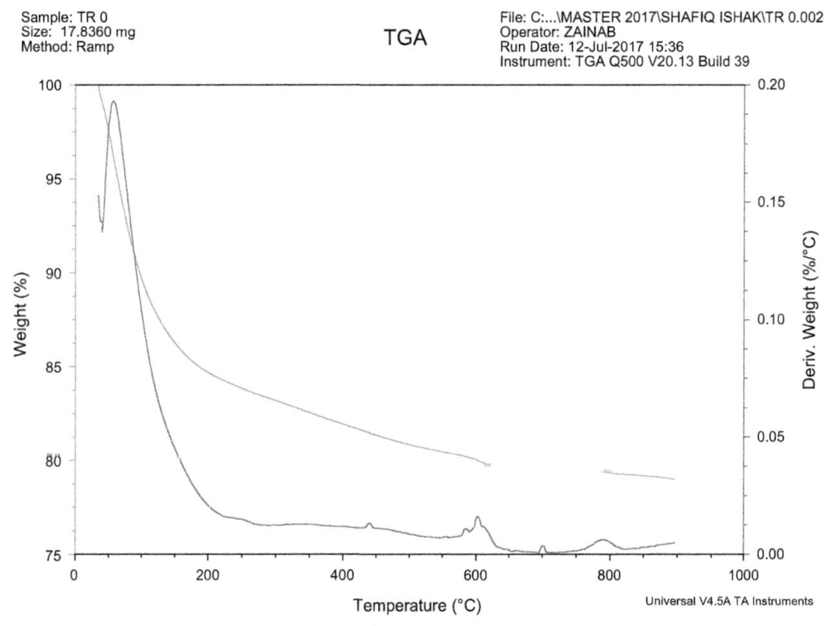

TGA/DTA Curves for the SCGPC Control Sample (Mix 1)

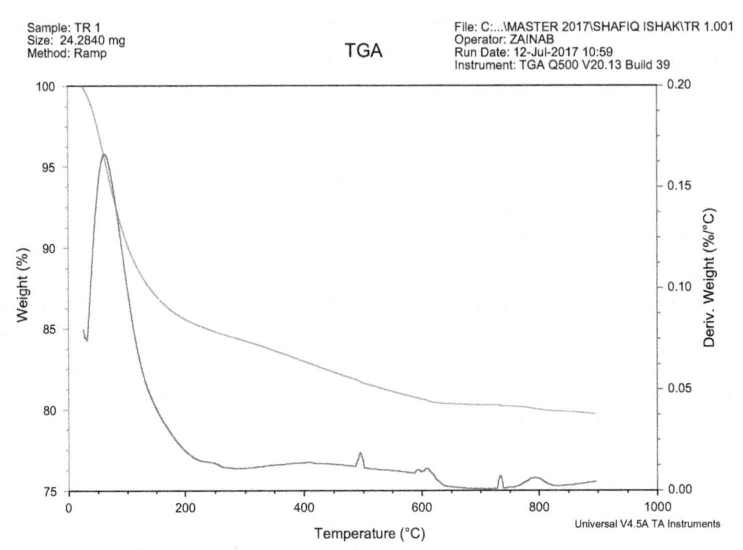

TGA/DTA Curves for the SCGPC Garnet Sample (Mix 2)

Chemical Compounds Present in GGBFS

Chemical Compounds	Composition (%)
SiO_2	33.80
Al_2O_3	13.68
Fe_2O_3	0.4
CaO	43.2
MgO	0.46
K_2O	0.21

Chemical Composition of Spent Garnet

Chemical Compounds	Weight % of Spent Garnet	Weight % of Sand
Fe_2O_3	43.06	0.7
SiO_2	33.76	96.4
$Al2O_3$	13.88	—
CaO	4.15	0.14
MgO	2.91	—
MnO	1.08	—
TiO_2	0.78	1.1
K_2O	0.14	—
P_2O_5	0.21	—
ZnO	0.06	—
Cr_2O_3	0.05	—

Properties	Spent Garnet	River Sand	Permissible Limits	Relevant Standards or Reference
Specific gravity	3.0	2.6	2.6–2.7	*PR
Fineness modulus	2.05	2.66	2.3–3.2	ASTM C33 (2003)
Hardness	7.5	6	1–10	Mohs scale
Bulk density (kg/m_3)	1922	1640	1300–1750	ASTM D 2003
Water absorption (%)	6	3	2–3	BS EN 1097-6 (2013)

References

ACI 211.4R-08 (2008). Guide for Selecting Proportions for High-Strength Concrete Using Portland Cement and Other Cementitious Materials. *ACI Manual of Concrete Practice*, part 1, p. 25. Farmington Hills, MI: American Concrete Institute.

ACI 318-02 (1995). Building Code Requirements for Structural Concrete and Commentary. Farmington Hills, MI: American Concrete Institute.

Aggarwal, Y., and Siddique, R. (2014). Microstructure and Properties of Concrete Using Bottom Ash and Waste Foundry Sand as Partial Replacement of Fine Aggregates. *Construction and Building Materials*, *54*, 210–223.

Akçaözoğlu, K. (2013). Microstructural Examination of Concrete Exposed to Elevated Temperature by Using Plane Polarized Transmitted Light Method. *Construction and Building Materials*, *48*, 772–779.

Al Qadi, A.N.S., and Al-Zaidyeen, S.M. (2014). Effect of Fibre Content and Specimen Shape on Residual Strength of Polypropylene Fibre Self-Compacting Concrete Exposed to Elevated Temperatures. *Journal of King Saud University: Engineering Sciences*, *26*(1), 33–39.

Ali, M.A., Krishnan, S., and Banerjee, D.C. (2001). Beach and Inland Heavy Mineral Sand Investigations and Deposits in India: An Overview. *Exploration and Research for Atomic Minerals*, *13*, 1–21.

Al-Jabri, K.S., Al-Saidy, A.H., and Taha, R. (2011). Effect of Copper Slag as a Fine Aggregate on the Properties of Cement Mortars and Concrete. *Construction and Building Materials*, *25*, 933–938.

Al-Saleh, S.A., and Al-Zaid, R.Z. (2006). Effects of Drying Conditions, Admixtures and Specimen Size on Shrinkage Strains. *Cement and Concrete Research*, *36*(10), 1985–1991.

Anastasiou, E., Georgiadis Filikas, K., and Stefanidou, M. (2014). Utilization of Fine Recycled Aggregates in Concrete with Fly Ash and Steel Slag. *Construction and Building Materials*, *50*, 154–161.

Ariffin, M.A.M., and Hussin, M.W. (2015). Chloride Resistance of Blended Ash Geopolymer Concrete. *Journal of Civil Engineering*, *6*(2), 23–33.

Arioz, O. (2007). Effects of Elevated Temperatures on Properties of Concrete. *Fire Safety Journal*, *42*(8), 516–522.

Bai, Y., Darcy, F., and Basheer, P.A.M. (2005). Strength and Drying Shrinkage Properties of Concrete Containing Furnace Bottom Ash as Fine Aggregate. *Construction and Building Materials*, *19*(9), 691–697.

Bakharev, T. (2005a). Geopolymeric Materials Prepared Using Class F Fly Ash and Elevated Temperature Curing. *Cement and Concrete Research*, *35*(6), 1224–1232.

Bakharev, T. (2005b) Resistance of Geopolymer Materials to Acid Attack. *Cement and Concrete Research*, *35*(4), 658–670.

Bakharev, T., Sanjayan, J.G., and Cheng, Y.B. (2003). Resistance of Alkali-Activated Slag Concrete to Acid Attack. *Cement and Concrete Research*, *33*(10), 1607–1611.

Barbosa, V.F., MacKenzie, K.J., and Thaumaturgo, C. (2000). Synthesis and Characterisation of Materials Based on Inorganic Polymers of Alumina and Silica: Sodium Polysialate Polymers. *International Journal of Inorganic Materials*, *2*(4), 309–317.

Basheer, P.A.M., Russell, D.P., and Rankin, G.I.B. (1999). 40 Design of Concrete to Resist Carbonation, (Vol. 1, pp. 423–435). Ottawa, ON, Canada: NRC Research Press.

Bassuoni, M.T., and Nehdi, M.L. (2007). Resistance of Self-Consolidating Concrete to Sulfuric Acid Attack with Consecutive pH Reduction. *Cement and Concrete Research*, *37*(7), 1070–1084.

Bera, A., Kumar, T., Ojha, K., and Mandal, A. (2013). Adsorption of Surfactants on Sand Surface in Enhanced Oil Recovery: Isotherms, Kinetics and Thermodynamic Studies. *Applied Surface Science*, *284*, 87–99.

Bissonnette, B., Pierre, P., and Pigeon, M. (1999). Influence of Key Parameters on Drying Shrinkage of Cementitious Materials. *Cement and Concrete Research*, *29*(10), 1655–1662.

Bondar, D., Lynsdale, C.J., Milestone, N., Hassani, N., and Ramezanianpour, A. (2011). Effect of Type, Form, and Dosage of Activators on Strength of Alkali-Activated Natural Pozzolans. *Cement and Concrete Research*, *33*, 251–260.

BS 812-103.1 (2011). *Methods for Determination of Particle Size Distribution —Sieve Tests.* London: British Standards Institution.

BS 882 (1992). *Specification for Aggregate from Natural Sources for Concrete.* London: BSI Standards Publication.

BS 882 (2011). *Specification for Aggregate from Natural Sources for Concrete.* London: British Standards Institution.

BS EN 1097-6 (2013). *Determination of Particle Density and Water Absorption.* London: British Standards Institution.

BS EN 12350-6 (2009). *Testing Fresh Concrete Density.* London: British Standards Institution.

BS EN 12390-7 (2009). *Density of Hardened Concrete.* London: British Standards Institution.

Cai, L., Ma, B., Li, X., Lv, Y., Liu, Z., and Jian, S. (2016). Mechanical and Hydration Characteristics of Autoclaved Aerated Concrete Containing Iron-Tailings: Effect of Content and Fineness. *Construction and Building Materials*, *128*, 361–372.

Castel, A.L., Cleary, J.D., and Pearson, C.E. (2010). Repeat Instability as the Basis for Human Diseases and as a Potential Target for Therapy. *Nature Reviews: Molecular Cell Biology*, *11*(3), 165–170.

Chan, D., and Sun, P.C. (2013). Effects of Fine Recycled Aggregate as Sand Replacement in Concrete. *HKIE Transactions*, *13*(4), 2–7.

Choi, Y.W., Kim, Y.J., Choi, O., Lee, K.M., and Lachemi, M. (2009). Utilization of Tailings from Tungsten Mine Waste as a Substitution Material for Cement. *Construction and Building Materials*, *23*(7), 2481–2486.

Cree, D., Green, M., and Noumowé, A. (2013). Residual Strength of Concrete Containing Recycled Materials after Exposure to Fire: A Review. *Construction and Building Materials*, *45*, 208–223.

Criado, M., Fernández-Jiménez, A., De La Torre, A.G., Aranda, M.A.G., and Palomo, A. (2007). An XRD Study of the Effect of the SiO 2/Na 2 O Ratio on the Alkali Activation of Fly Ash. *Cement and Concrete Research*, *37*(5), 671–679.

Das, B., Prakash, S., Reddy, P.S.R., and Misra, V.N. (2007). An Overview of Utilization of Slag and Sludge from Steel Industries. *Resources, Conservation and Recycling*, *50*(1), 40–57.

Davidovits, J. (1991). Geopolymers: Inorganic Polymeric New Materials. *Journal of Thermal Analysis and Calorimetry*, *37*(8), 1633–1656.

Davidovits, J. (1978). *J. of Applied Polymer Symposia* (IUPAC Symposium on Macromolecules, Stock-holm 1976, Topic III).

Davidovits, J. (1994). Recent Progresses in Concretes for Nuclear Waste and Uranium Waste Containment. *Concrete International*, *16*(12), 53–58.

Davidovits, J. (1999). Chemistry of Geopolymeric Systems Terminology. Proceedings of Geopolymer. International Conference, France, 1999.

Davidovits, J. (2002). Years of Successes and Failures in Geopolymer Applications. Market Trends and Potential Breakthroughs. *Keynote, Geopolymer 2002 Conference*. Vol. 28. Geopolymer Institute Saint-quentin (France), Melbourne (Australia).

Davidovits, J. (Ed.). (2005). *Geopolymer, Green Chemistry and Sustainable Development Solutions: Proceedings of the Geopolymer 2005 World Congress*. Saint-Quentin, France: Geopolymer Institute.

Demirel, B., and Keleştemur, O. (2010). Effect of Elevated Temperature on the Mechanical Properties of Concrete Produced with Finely Ground Pumice and Silica Fume. *Fire Safety Journal*, *45*(6–8), 385–391.

Devi, V.S., and Gnanavel, B.K. (2014). Properties of Concrete Manufactured Using Steel Slag. *Procedia Engineering*, *97*, 95–104.

Dinakar, P., Sethy, K.P., and Sahoo, U.C. (2013). Design of Self-Compacting Concrete with Ground Granulated Blast Furnace Slag. *Materials & Design*, *43*, 161–169.

Domone, P.L. (2006). Self-Compacting Concrete: An Analysis of 11 Years of Case Studies. *Cement and Concrete Composites*, *28*(2), 197–208.

Duan, P., Yan, C., Zhou, W., Luo, W., and Shen, C. (2015). An Investigation of the Microstructure and Durability of a Fluidized Bed Fly Ash–Metakaolin Geopolymer after Heat and Acid Exposure. *Materials & Design*, *74*, 125–137.

Duxson, P., Provis, J.L., Lukey, G.C., Mallicoat, S.W., Kriven, W.M., and Van Deventer, J.S. (2005). Understanding the Relationship Between Geopolymer Composition, Microstructure and Mechanical Properties. *Colloids and Surfaces A: Physicochemical and Engineering Aspects*, *269*(1–3), 47–58.

EFNARC, S. (2002). Guidelines for Self-Compacting Concrete.London: EFNARC.

Evangelista, L., and De Brito, J. (2010). Durability Performance of Concrete Made with Fine Recycled Concrete Aggregates. *Cement and Concrete Composites*, *32*(1), 9–14.

Evangelista, L., and De Brito, J. (2007). Mechanical Behaviour of Concrete Made with Fine Recycled Concrete Aggregates. *Cement and Concrete Composites*, *29*(5), 397–401.

Evans, J.G., Spiess, P.E., Kamat, A.M., Wood, C.G., Hernandez, M., Pettaway, C.A., and Pisters, L.L. (2006). Chylous Ascites after Post-Chemotherapy Retroperitoneal Lymph Node Dissection: Review of the MD Anderson Experience. *The Journal of Urology*, *176*(4), 1463–1467.

Fares, H., Remond, S., Noumowe, A., and Cousture, A. (2010). High Temperature Behaviour of Self-Consolidating Concrete: Microstructure and Physicochemical Properties. *Cement and Concrete Research*, *40*(3), 488–496.

Feng, X.X., Xi, X.L., Cai, J.W., Chai, H.J., and Song, Y.Z. (2011). Investigation of Drying Shrinkage of Concrete Prepared with Iron Mine Tailings. *Key Engineering Materials*, *477*, 37–41.

Fernández-Jiménez, A., and Palomo, A. (2005). Mid-infrared Spectroscopic Studies of Alkali-Activated Fly Ash Structure. *Microporous and Mesoporous Materials*, *86*(1–3), 207–214.

Flatt, R.J., Roussel, N., and Cheeseman, C.R. (2012). Concrete: An Eco Material that Needs to Be Improved. *Journal of the European Ceramic Society*, *32*(11), 2787–2798.

Garnett, T., Appleby, M.C., Balmford, A., Bateman, I.J., Benton, T.G., and Bloomer, H.M. (2013). Sustainable Intensification in Agriculture: Premises and Policies. *Science*, *341*(6141), 33–34.

Georgali, B., and Tsakiridis, P.E. (2005). Microstructure of Fire-Damaged Concrete. A Case Study. *Cement and Concrete Composites*, *27*(2), 255–259.

Giergiczny, Z., and Krol, A. (2008). Immobilization of Heavy Metals (Pb, Cu, Cr, Zn, Cd, Mn) in the Mineral Additions Containing Concrete Composites. *Hazard Materials*, *160*(2–3), 247–255.

Gonen, T., and Yazicioglu, S. (2007). The Influence of Compaction Pores on Sorptivity and Carbonation of Concrete. *Construction and Building Materials*, *21*(5), 1040–1045.

Gorhan, G., and Kurklu, G. (2014). The Influence of the NaOH Solution on the Properties of the Fly Ash–Based Geopolymer Mortar Cured at Different Temperatures. *Composites Part B: Engineering, 58*, 371–377.

Gorrill, L. 2003. Global Garnet Market Review. *Mineral Price Watch, 97*(January), 7–10.

Gourley, J.J., and Calvert, C.M. (2003). Automated Detection of the Bright Band Using WSR-88D Data. *Weather and Forecasting, 18*(4), 585–599.

Goyal, S., Singh, K., Hussain, A., and Singh, P.R. (2015). Study on Partial Replacement of Sand with Iron Ore Tailing on Compressive Strength of Concrete. *International Journal of Research in Engineering & Advanced Technology, 3*(2).

Haranki, B. 2009. Strength, Modulus of Elasticity, Creep and Shrinkage of Concrete Used in Florida (MA thesis). Gainesville, University of Florida.

Hardjito, D., and Rangan, B.V. (2005). *Development and Properties of Low-Calcium Fly Ash–Based Geopolymer Concrete.*

Hardjito, D., Wallah, S.E., Sumajouw, D.M.J., and Rangan, B.V. (2004). On the Development of Fly Ash–Based Geopolymer Concrete. *ACI Material Journal, 101*(6), 467–472.

Harris, P. (2000). At the Cutting Edge: Abrasives and Their Markets. *Industrial Minerals, 388*(January), 19–27.

Helene, Paulo, Pereira, M. F., and Castro, P.. (2004). Performance of a 40-Year-Old Concrete Bridge with Embedded, Prestressed Galvanized Strands. *Materials Performance, 43*(10), 42–45.

Holt, E., and Leivo, M. (2004). Cracking Risks Associated with Early Age Shrinkage. *Cement and Concrete Composites, 26*(5), 521–530.

Huang, X., Ranade, R., and Li, V.C. (2013). Feasibility Study of Developing Green ECC Using Iron Ore Tailings Powder as Cement Replacement. *Journal of Materials In Civil Engineering, 25*(7), 923–931.

Ismail, M., Elgelany Ismail, M., and Muhammad, B. (2011). Influence of Elevated Temperatures on Physical and Compressive Strength Properties of Concrete Containing Palm Oil Fuel Ash. *Construction and Building Materials, 25*(5), 2358–2364.

Ismail, S., Hoe, K.W., and Ramli, M. (2013). Sustainable Aggregates: The Potential and Challenge for Natural Resources Conservation. *Procedia: Social and Behavioral Sciences, 101*, 100–109.

Jeffrey, K. (2006). Classification of Industrial Minerals and Rocks. *Industrial Minerals & Rocks: Commodities, Markets, and Uses (7th ed).* Littleton, CO: SME, 7–11.

Jiminez, A.M.F., Lachowski, E.E., Palomo, A., and Macphee, D.E. (2004). Microstructural Characterisation of Alkali-Activated PFA Matrices for Waste Immobilisation. *Cement and Concrete Composites, 26*(8), 1001–1006.

Kamada, H., Okamura, N., Satake, M., Harada, H., and Shimomura, K. (1986). Alkaloid Production by Hairy Root Cultures in Atropa Belladonna. *Plant Cell Reports, 5*(4), 239–242.

Kang, H.Z., Jia, K.W., and Yao, L. (2011). Experimental Study on Properties of Concrete Mixed with Ferrous Mill Tailing. *Applied Mechanics and Materials, 148–149*, 904–907.

Kaufhold, S., Hein, M., Dohrmann, R., and Ufer, K. (2012). Quantification of the Mineralogical Composition of Clays Using FTIR Spectroscopy. *Vibrational Spectroscopy, 59*, 29–39.

Khatib, J.M., and Ellis, D.J. (2001). Mechanical Properties of Concrete Containing Foundry Sand. *Special Publication, 200*, 733–748.

Khunthongkeaw, J., Tangtermsirikul, S., and Leelawat, T. (2006). A Study on Carbonation Depth Prediction for Fly Ash Concrete. *Construction and Building Materials*, 20(9), 744–753.

Kodur, V.K.R., Cheng, F.P., Wang, T.C., and Sultan, M.A. (2003). Effect of Strength and Fiber Reinforcement on Fire Resistance of High-Strength Concrete Columns. *Journal of Structural Engineering, ASCE Library, 129*, 253–259.

Kondolf, G.M., Angermeier, P.L., Cummins, K., Dunne, T., Healey, M., Kimmerer, W., and Reed, D.J. (2008). Projecting Cumulative Benefits of Multiple River Restoration Projects: An Example from the Sacramento–San Joaquin River System in California. *Environmental Management*, *42*(6), 933–945.

Kong, F.R., Pan, L.S., Wang, C.M., and Xu, N. (2016). Effects of Polycarboxylate Superplasticizers with Different Molecular Structure on the Hydration Behavior of Cement Paste. *Construction and Building Materials*, *105*, 545–553.

Konrad-Schmolke, M., Babist, J., Handy, M.R., and O'Brien, P.J. (2006). The Physico-chemical Properties of a Subducted Slab from Garnet Zonation Patterns (Sesia Zone, Western Alps). *Journal of Petrology*, *47*(11), 2123–2148.

Kori, E., and Mathada, H. (2012). An Assessment of Environmental Impacts of Sand and Gravel Mining in Nzhelele Valley, Limpopo Province, South Africa. In *Proceedings of the Third International Conference on Biology, Environment and Chemistry*, pp. 137–141.

Kosmatka, S.H., Panarese, W.C., and Kerkhoff, B. (2002). *Design and Control of Concrete Mixtures*, (Vol. 54, pp. 1077–1083). Skokie, IL: Portland Cement Association.

Kou, S.C., and Poon, C.S. (2009). Properties of Self-Compacting Concrete Prepared with Coarse and Fine Recycled Concrete Aggregates. *Cement and Concrete Composites*, *31*(9), 622–627.

Kore, S.D., and Vyas, A.K.. (2017). Impact of Fire on Mechanical Properties of Concrete Containing Marble Waste. *Journal of King Saud University: Engineering Sciences*, (in press).

Krishna, K.M., Reddy, K.S.N., Sekhar, C.R., Naidu, K.B., Rao, P.G., and Reddy, G.V.R. (2016). Heavy Mineral Studies on Late Quaternary Red Sediments of Bhimunipatnam, Andhra Pradesh, East Coast of India. *Journal of the Geological Society of India*, *88*(5), 637–647.

Kristof, J., Frost, R.L., Felinger, A., and Mink, J. (1997). FTIR Spectroscopic Study of Intercalated Kaolinite. *Journal of Molecular Structure*, *411*, 119–122.

Łaźniewska-Piekarczyk, B. (2014). The Methodology for Assessing the Impact of New Generation Superplasticizers on Air Content in Self-Compacting Concrete. *Construction and Building Materials*, *53*, 488–502.

Li, G., Chen, Y.J., and Di, H. (2013). The Study on Used Properties of Mine Tailings Sand. *Advanced Materials Research*, *859*, 87–90.

Lin, K.L., Wang, K.S., Tzeng, B.Y., and Lin, C.Y. (2003). The Reuse of Municipal Solid Waste Incinerator Fly Ash Slag as a Cement Substitute. *Resources, Conservation and Recycling*, *39*(4), 315–324.

Lin, Y., Hsiao, C., Yang, H., and Lin, Y.F. (2011). The Effect of Post-Fire-Curing on Strength–Velocity Relationship for Nondestructive Assessment of Fire-Damaged Concrete Strength. *Fire Safety Journal*, *46*(4), 178–185.

Lindtner, S., Hertz, G.D., and Dourish, P. (2014). Emerging Sites of HCI Innovation: Hackerspaces, Hardware Startups and Incubators. In *Proceedings of the SIGCHI Conference on Human Factors in Computing Systems*, pp. 439–448.

Lottermoser, B.G. (2011). Recycling, Reuse and Rehabilitation of Mine Wastes. *Elements*, *7*(6), 405–410.

Maekawa, K., Takemura, J.I., Irawan, P., and Irie, M.A. (1993). Triaxial Elasto-Plastic and Fracture Model for Concrete. *Doboku Gakkai Ronbunshu*, *460*, 131–138.

Meenakshi, S.S., and Ilangovan, R. (2011). Performance of Copper Slag and Ferrous Slag as Partial Replacement of Sand in Concrete. *International Journal of Civil and Structural Engineering*, *1*(4), 918–927.

Mehta, P.K. (1986). Effect of Fly Ash Composition on Sulfate Resistance of Cement. *Journal Proceedings*, *83*(6), 994–1000.

Mehta, S.R., Yusuf, S., Peters, R.J., Bertrand, M.E., Lewis, B.S., Natarajan, M.K., and Copland, I. (2001). Effects of Pretreatment with Clopidogrel and Aspirin Followed by Long-Term Therapy in Patients Undergoing Percutaneous Coronary Intervention: The PCI-CURE Study. *The Lancet*, *358*(9281), 527–533.

Memon, F.A., Nuruddin, M.F., Demie, S., and Shafiq, N. (2012). Effect of Superplasticizer and Extra Water on Workability and Compressive Strength of Self-Compacting Geopolymer Concrete. *Research Journal of Applied Sciences, Engineering and Technology*, *4*(5), 407–414.

Meyer, C. (2009). The Greening of the Concrete Industry. *Cement and Concrete Composites*, *31*(8), 601–605.

Miranda, J.C. (2005). On Tuned Mass Dampers for Reducing the Seismic Response of Structures. *Earthquake Engineering & Structural Dynamics*, *34*(7), 847–865.

Monteny, J., Vincke, E., Beeldens, A., De Belie, N., Taerwe, L., Van Gemert, D., and Verstraete, W. (2000). Chemical, Microbiological, and In Situ Test Methods for Biogenic Sulfuric Acid Corrosion of Concrete. *Cement and Concrete Research*, *30*(4), 623–634.

Morsya, M.M., Shebla, S.S., and Rashadb, A.M. (2008). Effect of Fire on Microstructure and Mechanical Properties of Blended Cement Pastes Containing Metakaolin and Silica Fume. *Asian Journal of Civil Engineering*, *9*(2), 93–105.

Nagral, M.R., Ostwal, T., and Chitawadagi, M.V. (2014). Effect of Curing Temperature and Curing Hours on the Properties of Geo-polymer Concrete. *International Journal of Computational Engineering Research*, *4*, 2250–3005.

Nematzadeh, M., and Fallah-Valukolaee, S. (2017). Erosion Resistance of High-Strength Concrete Containing Forta-Ferro Fibers against Sulfuric Acid Attack with an Optimum Design. *Construction and Building Materials*, *154*, 675–686.

Neville, A.M. (2011). *Properties of Concrete.* London: Pearson Education.

Neville, A.M., and Brooks, J.J. (2010). *Concrete Technology.* Essex, UK: Pearson Education.

Nikolov, A., and Rostovsky, I., (2014) *Structural Study of Geopolymers Based on Natural Zeolite and Sodium Silicate after Thermal Treatment of up to 1000°C*, Annual of the University of Architecture, Civil Engineering and Geodesy, Sofia, Fascicule VIII-A, Scientific Research I, 2014–2015, University of Architecture, Sofia, Bulgaria, 2014, 1310-814.

Nuruddin, M.F., Kusbiantoro, A., Qazi, S., and Shafiq, N. (2011). Compressive Strength and Interfacial Transition Zone Characteristic of Geopolymer Concrete with Different Cast In-Situ Curing Condition, World Academy of Science, Engineering and Technology (WASET) Conference, Dubai, UAE, 2011, pp. 25–28.

Nuruddin, M.F., and Memon, F.A. (2015). Properties of Self-Compacting Geopolymer Concrete. *Materials Science Forum*, *803*, 99–109.

Oh, J.E., Moon, J., Mancio, M., Clark, S.M, and Monterio, P.J.M. (2011). Bulk Modulus of Basic Sodalite, $Na_8[AlSiO_4]6(OH)_2 2H_2O$, a Possible Zeolitic Precursor in Coal-Fly-Ash-Based Geopolymers. *Cement and Concrete Research*, *41*(1), 107–112.

Olson, D.W. (2001). Garnet, Industrial. *U.S. Geological Survey Minerals Yearbook*, 30–31.

Olson, D.W. (2005). Garnet (Industrial). *U.S. Geological Survey Mineral Commodity Summaries*, 66–67.

Ozawa, K. (1989). High-Performance Concrete based on the Durability Design of Concrete Structures. *Proc. of the Second East Asia-Pacific Conference on Structural Engineering and Construction.*

Palomo, A., Grutzeck, M.W., and Blanco, M.T. (1999). Alkali-Activated Fly Ashes: A Cement for the Future. *Cement and Concrete Research*, *29*(8), 1323–1329.

Panda, C.R., Mishra, K.K., Panda, K.C., Nayak, B.D., and Nayak, B.B. (2013). Environmental and Technical Assessment of Ferrochrome Slag as Concrete Aggregate Material. *Construction and Building Materials*, *49*, 262–271.

Panias, D., Giannopoulou, I.P., and Perraki, T. (2007). Effect of Synthesis Parameters on the Mechanical Properties of Fly Ash–Based Geopolymers. *Colloids Surf A*, *301*(1–3), 246–254.

Park, C.K. (2000). Hydration and Solidification of Hazardous Wastes Containing Heavy Metals Using Modified Cementitious Materials. *Cement and Concrete Research*, *30*(3), 429–435.

Park, S.B., Lee, B.C., and Kim, J.H. (2004). Studies on Mechanical Properties of Concrete Containing Waste Glass Aggregate. *Cement and Concrete Research*, *34*(12), 2181–2189.

Part, W.K., Ramli, M., and Cheah, C.B. (2015). An Overview on the Influence of Various Factors on the Properties of Geopolymer Concrete Derived from Industrial By-Products. *Construction and Building Materials*, *77*, 370–395.

Parthiban, K., Saravanarajamohan, K., Shobana, S., and Bhaskar, A.A. (2013). Effect of Replacement of Slag on the Mechanical Properties of Fly Ash Based Geopolymer Concrete. *International Journal of Engineering and Technology*, *5*(3), 2555–2559.

Pettijohn, F.J., Potter, P.E., and Siever, R. (2012). *Sand and Sandstone. Progress in Photovoltaics: Research and Applications*, *8*(1), 61–76.

Prabhu, G.G., Hyun, J.H., and Kim, Y.Y. (2014). Effects of Foundry Sand as a Fine Aggregate in Concrete Production. *Construction and Building Materials*, *70*, 514–521.

Provis, J.L., Duxson, P., and van Deventer, J.S. (2010). The Role of Particle Technology in Developing Sustainable Construction Materials. *Advanced Powder Technology*, *21*(1), 2–7.

Rajamanickam, G.V., Chandrasekar, N., Angusamy, N., and Loveson, V.J. (2004). Status of Beach Placer Mineral Exploration in India. *Sustainable Development of Coastal Placer Minerals*, pp. 9–21. Loveson, V.J., and Misra, D.D. (Eds). New Delhi, India: Allied.

Rangan, B.V. (2008). Fly Ash–Based Geopolymer Concrete. Report GC4, Curtin University, Perth, Australia, 1–44.

Rao, G.A. (2001). Long-Term Drying Shrinkage of Mortar — Influence of Silica Fume and Size of Fine Aggregate. *Cement and Concrete Research*, *31*(2), 171–175.

Rao, A., Jha, K.N., and Misra, S. (2007). Use of Aggregates from Recycled Construction and Demolition Waste in Concrete. *Resources, Conservation and Recycling*, *50*(1), 71–81.

Richardson, A.E., Coventry, K., and Graham, S. (2009). Concrete Manufacture with Un-graded Recycled Aggregates. *Structural Survey*, *27*(1), 62–70.

Rodina, I.S., Kravtsova, A.N., Soldatov, A.V., Yalovega, G.E., Popov, Y.V., and Boyko, N.I. (2013). X-Ray Spectroscopic Identification of Garnet from the Placer Deposits of the Taman Peninsula. *Optics and Spectroscopy*, *115*(6), 858–862.

Roskill Information Services. (2000). *The Economics of Garnet*, 3rd Ed. London: Roskill Information Services.

Ryan, J.M., and Harris, S.P. (2000). Using State of the Art Blast Modeling Software to Assist the Excavation of the Yucca Mountain Nuclear Waste Repository. In *High-Tech Seminar: State-of-the-Art Blasting Technology Instrumentation and Explosives Applications*, pp. 407–423.

Sagoe, C.K.K., Brown, T., and Taylor, A.H. 2002. Durability and Performance Characteristics of Recycled Aggregate Concrete. *CSIRO Building Construction and Engineering*, Victoria, Australia.

Saifuddin, K.P., and Purohit, B.M. (2014). Effects of Superplasticizer on Self-Compacting Geopolymer Concrete Using Fly Ash and Ground Granulated Blast Furnace Slag. *Journal of International Academic Research for Multidisciplinary*, *2*(3), 2320–5083.

Sanni, S.H., and Khadiranaikar, R.B. (2013). Performance of Alkaline Solutions on Grades of Geopolymer Concrete. *International Journal of Research in Engineering and Technology*, *2*(11), 366–371.

Sashidhar, C., Jawahar, J.G., Neelima, C., and Kumar, D.P. (2015). Fresh and Strength Properties of Self Compacting Geopolymer Concrete Using Manufactured Sand. *International Journal of ChemTech Research*, 8(7), 183–190.

Savva, A., Manita, P., and Sideris, K.K. (2005). Influence of Elevated Temperatures on the Mechanical Properties of Blended Cement Concretes Prepared with Limestone and Siliceous Aggregates. *Cement and Concrete Composites*, *27*(2), 239–248.

Shafiq, I., Azreen, M., and Hussin, M.W. (2017). Sulphuric Acid Resistant of Self-Compacted Geopolymer Concrete Containing Slag and Ceramic Waste. *MATEC Web of Conferences*, *97*, 01102.

Shettima, A.U., et al. (2016). Evaluation of Iron Ore Tailings as Replacement for Fine Aggregate in Concrete. *Construction and Building Materials*, *120*, 72–79.

Shi, H.S., Xu, B.W., and Zhou, X.C. (2009). Influence of Mineral Admixtures on Compressive Strength, Gas Permeability and Carbonation of High Performance Concrete. *Construction and Building Materials*, *23*(5), 1980–1985.

Siddique, R., Schutter, G. D., and Noumowe, A. (2009). Effect of Used-Foundry Sand on the Mechanical Properties of Concrete. *Construction and Building Materials* *23*(2), 976–980.

Siddique, R., Aggarwal, Y., Aggarwal, P., Kadri, E.H., and Bennacer, R. (2011). Strength, Durability, and Micro-structural Properties of Concrete Made with Used-Foundry Sand (UFS). *Construction and Building Materials*, *25*(4), 1916–1925.

Singh, M., and Siddique, R. (2013). Effect of Coal Bottom Ash as Partial Replacement of Sand on Properties of Concrete. *Resources, Conservation and Recycling*, *72*, 20–32.

Singh, M., Siddique, R., Ait-Mokhtar, K., and Belarbi, R. (2016). Durability Properties of Concrete Made with High Volumes of Low-Calcium Coal Bottom Ash as a Replacement of Two Types of Sand. *Journal of Materials in Civil Engineering*, *28*(4), 04015175.

Somna, K., Jaturapitakkul, C., Kajitvichyanukul, P., and Chindaprasirt, P. (2011). NaOH-Activated Ground Fly Ash Geopolymer Cured at Ambient Temperature. *Fuel*, *90*(6) , 2118–2124.

Sreenivasulu, C., Guru Jawahar, J., Vijaya Sekhar Reddy, M., and Pavan Kumar, D. (2016). Effect of Fine Aggregate Blending on Short-Term Mechanical Properties of Geopolymer Concrete. *Asian Journal of Civil Engineering*, *17*(5), 537–550.

Swanepoel, J.C., and, Strydom, C.A. (2002). Utilisation of Fly Ash in a Geopolymeric Material. *Applied Geochemistry*, *17*(8), 1143–1148.

Tahri, W., Abdollahnejad, Z., Mendes, J., Pacheco-Torgal, F., and de Aguiar, J.B. (2017). Cost Efficiency and Resistance to Chemical Attack of a Fly Ash Geopolymeric Mortar versus Epoxy Resin and Acrylic Paint Coatings. *European Journal of Environmental and Civil Engineering*, *21*(5), 555–571.

Temuujin, J., van Riessen, A., and MacKenzie, K.J.D. (2010). Preparation and Characterisation of Fly Ash Based Geopolymer Mortars. *Construction and Building Materials*, *24*(10), 1906–1910.

Teng, S., Lim, T.Y.D., and Divsholi, B.S. (2013). Durability and Mechanical Properties of High Strength Concrete Incorporating Ultra Fine Ground Granulated Blast-Furnace Slag. *Construction and Building Materials*, *40*, 875–881.

Teixeira-Pinto, A., Fernandes, P., and Jalali, S. (2002). Geopolymer Manufacture and Application-Main Problems When Using Concrete Technology. *Geopolymers 2002 International Conference,* Melbourne, Australia, Siloxo Pty. Ltd.

Thampi, T., Sreevidya, V., and Venkatasubramani, R. (2014). Strength Studies on Geopolymer Mortar for Ferro Geopolymer Water Tank. *International Journal of Advanced Structures and Geotechnical Engineering*, *3*, 102–105.

Tripathi, B., Misra, A., and Chaudhary, S., (2013). Strength and Abrasion Characteristics of ISF Slag Concrete. *Journal of Materials in Civil Engineering*, *25*, 1611–1618.

Ugama, T.I. and Ejeh, S.P. (2014). Iron Ore Tailing as Fine Aggregate in Mortar Used for Masonry. *International Journal of Advances in Engineering & Technology*, *7*(4), 1170–1178.

U.S EPA. (2000). Sources of dioxin-like compounds in the United States. Draft exposure and human health reassessment of 2, 3, 7, 8-tetrachlorodibenzo-p-dioxin (TCDD) and related compounds. Rep. No. EPA/600/p-00/001Bb, Washington, DC.

Valcuende, M., and Parra, C. (2010). Natural Carbonation of Self-Compacting Concretes. *Construction and Building Materials*, *24*(5), 848–853.

Van Jaarsveld, J.G.S., Van Deventer, J.S.J., and Lukey, G.C. (2003). The Characterisation of Source Materials in Fly Ash–Based Geopolymers. *Materials Letters*, *57*(7), 1272–1280.

Xu, H., and Van Deventer, J.S.J. (2000). The Geopolymerisation of Alumino-silicate Minerals. *International Journal of Mineral Processing*, *59*(3), 247–266.

Yehia, S., Helal, K., Abusharkh, A., Zaher, A., and Istaitiyeh, H. (2015). Strength and Durability Evaluation of Recycled Aggregate Concrete. *International Journal of Concrete Structures and Materials*, *9*(2), 219–239.

Yongde, Li and Yao Sun. (2000). Preliminary Study on Combined-Alkali–Slag Paste Materials. *Cement and Concrete Research*, *30*(6), 963–966.

Zaharieva, R., Buyle-Bodin, F., and Wirquin, E. (2004). Frost Resistance of Recycled Aggregate Concrete. *Cement and Concrete Research*, *34*(10), 1927–1932.

Žarnić, R., Gostič, S., Crewe, A.J., and Taylor, C.A. (2001). Shaking Table Tests of 1: 4 Reduced-Scale Models of Masonry Infilled Reinforced Concrete Frame Buildings. *Earthquake Engineering & Structural Dynamics*, *30*(6), 819–834.

Zhang, G.D., Zhang, X.Z., Zhou, Z.H., and Cheng, X. (2013). Preparation and Properties of Concrete Containing Iron Tailings/Manufactured Sand as Fine Aggregate. *Advanced Materials Research*, *838–841*, 152–155.

Zhao, H., Sun, W., Wu, X., and Gao, B. (2015). The Properties of the Self-Compacting Concrete with Fly Ash and Ground Granulated Blast Furnace Slag Mineral Admixtures. *Journal of Cleaner Production*, *95*, 66–74.

Zhao, H., Sun, W., Wu, X., and Gao, B. (2017). The Effect of the Material Factors on the Concrete Resistance against Carbonation. *KSCE Journal of Civil Engineering*, *53*, 1–10.

Zhao, S., Fan, J., and Sun, W. (2014). Utilization of Iron Ore Tailings as Fine Aggregate in Ultra-High-Performance Concrete. *Construction and Building Materials*, *50*, 540–548.

Zhou, X.M., Slater, J.R., Wavell, S.E., and Oladiran, O. (2012). Effects of PFA and GGBS on Early-Ages Engineering Properties of Portland Cement Systems. *Journal of Advanced Concrete Technology*, *10*, 74–85.

Index

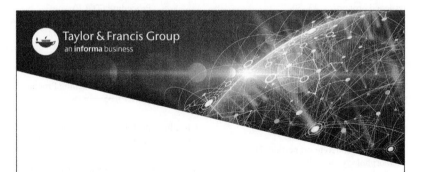